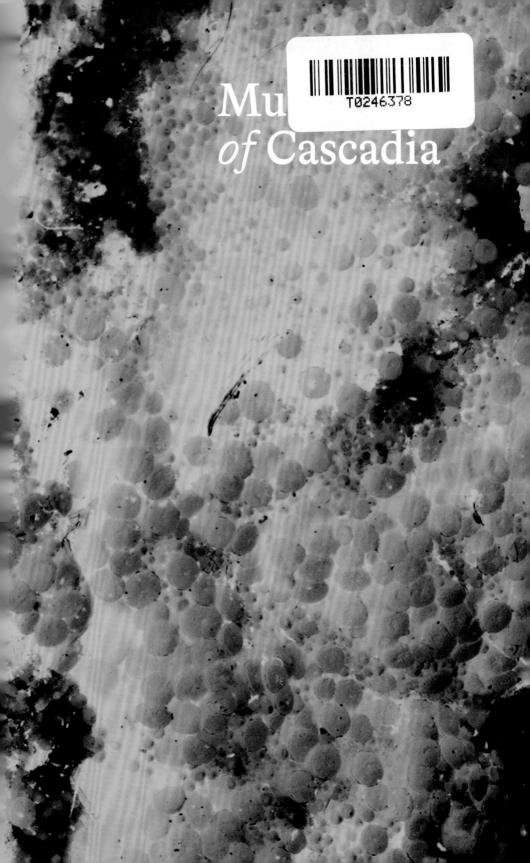

Mu
of Cascadia

Mushrooms
of Cascadia

An Illustrated Key to the Fungi of the Pacific Northwest
Second Edition

MICHAEL BEUG
PREFACE BY PAUL STAMETS

TEN SPEED PRESS
California | New York

Dedication

This book is dedicated to the memory of Kit Scates Barnhart and her husband, Harley Barnhart. In 1974, along with Dr. Daniel Stuntz, Kit founded the Pacific Northwest Key Council, an organization dedicated to writing keys for mushroom identification. My membership in the Key Council began in 1975 and gave me the chance to work with and learn from all the serious amateur and professional mycologists in the Pacific Northwest. Dr. Stuntz generously shared his vast mycological knowledge with all of us, and I was able to spend my first two sabbatical leaves working in his laboratory. Harley Barnhart regularly critiqued my photography. And Kit and Harley left the best of their vast photo collection to me after their passing. I have drawn heavily on their photos in creating this work.

Key Council members, now numbering more than sixty, were all dedicated to writing dichotomous keys, which are now available online for free at svims.ca/council/keys.htm. In 1995, Ian Gibson had the idea of creating a free computer program, now called MycoMatch, which was the inspiration for this book.

CONTENTS

Jim Gouin with a rare example of *Bridgeoporus nobilissimus*, the largest polypore of all, known only from a few old-growth forest sites in Cascadia (SEE KEY LEAD 91C [PAGE 113]).

Preface

Dr. Michael Beug's *Mushrooms of Cascadia* is a triumphant achievement in the annals of mycology. His illustrated key helps readers logically navigate through the often-bewildering array of mushrooms, choosing features that most relevantly separate one cluster of species from another. I have known Dr. Beug all of my adult life and watched his passion for—and compassion toward—the field of mycology; *Mushrooms of Cascadia* is a rare accomplishment.

Too many guides show just a few species over a broad geographical range, leaving out many species and sowing confusion. This book fills an important gap. Dr. Beug expertly guides you on an academic field trip to identify mushrooms safely. I applaud and highly recommend *Mushrooms of Cascadia* as the best field guide to the mysterious and magical mushrooms of the Pacific Northwest *and beyond*.

Well done, Professor!

—*Paul Stamets, June 2023*

Acknowledgments

My first edition of this book was made possible by Britt Bunyard, the owner of Fungi Press, and Paul Stamets. Technical readers for the first edition were Britt Bunyard, Paul Przybylowicz, Ian Gibson, and Danny Miller; Danny, Ian, and Kent Brothers did a technical read for this second edition. Important departed Key Council members include Ben Woo, Oluna Ceska, Margaret Dilly, Judy Roger, and Lorelei Norvell. Alissa Allen, Steve Trudell, Jim Ginns, Cathy Cripps, Fred Rhoades, Paul Kroeger, Dick Bishop, Jan Lindgren, Buck McAdoo, Dennis Oliver, Christine Roberts, Drew Parker, Joe Ammirati, and Jim Trappe all shaped this work. I owe much to my favorite teaching partner, Paul Przybylowicz, and to my many students, especially Helen Lau, Nicole Hynson, and Thom O'Dell. My wife, Frances Ann Beug, put up with my years spent writing this book.

Thanks to Jan Hammond, the production editor of the first edition. Julie Bennett of Ten Speed Press took me under her wing for this edition. I also wish to thank Annie Marino, Ashley Pierce, Mari Gill, Serena Sigona, and Stephanie Davis of Ten Speed Press.

Introduction

This story begins in the spring of 1968. I was a chemistry graduate student at the University of Washington and recently married to my life partner, Ann, then a fellow graduate student in chemistry. We were given a small bag of freshly picked morels that we sautéed for dinner. The flavor was rich and reminded me of a fine, tender steak. That fall, we took a non-credit mushroom* identification class taught by Dr. Daniel Stuntz at the University of Washington. The first Pacific Northwest mushroom clubs were just being formed, and only one small book was available for identifying Pacific Northwest mushrooms: Margaret McKenny's *The Savory Wild Mushroom.* I wanted to know all there was to know about mushrooms, and I am still learning fifty-five years later.

In the fall of 1972, I began my thirty-two years on the faculty of The Evergreen State College in Olympia, Washington. My initial assignment was to work with independent-study students. A small group of students wanted to learn about mushrooms. Olympia had been the home of Margaret McKenny, who had died

*I use the terms "fungi" and "mushrooms" interchangeably to apply to all macroscopic fungi even though, technically, mushrooms are only the fungi with a cap and stipe (or stem).

a few years earlier. She was an Audubon member and had started the tradition of a mushroom show every fall. It was at the Audubon Society Mushroom Show that I met Dr. Alexander H. Smith from the University of Michigan and Kit Scates from Post Falls, Idaho, both there to help with the show. Dr. Smith was then the leading North American mycologist, and he would return often to visit Evergreen and observe the amazing scanning electron microscope work that my student Paul Stamets was engaged in. (Dr. Smith did not have access to an SEM at the University of Michigan.) Kit Scates (later Kit Scates Barnhart) was an English teacher who was fascinated by fungi but frustrated by the lack of mushroom identification resources. Upon learning that Dr. Smith often spent his summers working at the Priest River Experimental Forest in Northern Idaho, she gathered all the different mushrooms she could find and drove north to meet him and learn about the mushrooms she was finding. It was the beginning of a lifelong association. At Dr. Smith's suggestion, Kit took on the study of coral mushrooms, especially the genus *Ramaria*. Kit became my mentor, and I inherited her fascination with coral mushrooms and indeed all fungi. My mushroom endeavors suddenly became serious.

I have since hunted and photographed mushrooms from Anchorage to San Diego to the Rockies, as well as occasionally on the Eastern Seaboard from Newfoundland to Georgia and the upper Midwest. My photo collection has shaped this book. Cascadia alone is an area with more than 4,500 named species and at least 12,000 more species still unnamed. Some are seen reliably every year (and are the focus of this book), while most are rarely seen. Many of these same species, or sister species with the same properties, are also found in the Midwest and the East, as well as in Europe, Eurasia, and Asia, and thus you can use this book to help identify choice edibles and safely steer away from potentially deadly species in all temperate ecosystems. While the exact chanterelle species, oyster species, *Hericium* species, matsutake, or truffle may differ in various parts of the world—and there will be differences in flavor; you will learn to recognize the genera—and thus what is edible and what is a dangerous poisonous species. If you identify a mushroom as the Alice in Wonderland mushroom, *Amanita muscaria*, you may in fact have a look-alike if you are not in Europe, but it will have the same toxins. The *Amanita* species known as destroying angels and death caps may be genetically slightly different in various parts of the world, but they are all deadly.

Incidentally, names of fungi (and, indeed, all living organisms) have been slowly evolving over at least the past several hundred years. However, recent advances in DNA technology have greatly accelerated the rate at which organisms are reclassified. Even popular sources for learning current "correct" fungal names, such as *Index Fungorum* and *MycoBank*, are struggling to get ahead of the current deluge of new names and may disagree with one another. If you use this book for scientific purposes, be prepared for many more changes. In fact, already much has changed since I wrote the first edition.

Throughout this book, I have indicated where a name change may already be required by using single quotation marks (for example, *'Mycena' acicula* means that a new genus is needed for the species mentioned, and *Mycena 'amicta'* means that the genus is not likely to need changing but the species epithet may change). When a specific new name has already been suggested, but not yet published, I have added n.p. after the name (for example *Amanita lindgreniana n.p.*). When there are several very similar species and some are unnamed, I have used "group" at the end of the species name. Completely unanticipated name changes may also arise. I will post updates and new species that I discover on my website, www.mushroomsofcascadia.com, and I post all new discoveries on iNaturalist under the name organicgardner44.

CHAPTER 1

Collecting and Identifying Mushrooms

Although the oak woodlands near my home have significant numbers of fleshy mushrooms only in November (and sometimes into December), my pursuit of mushrooms is year-round. The Columbia River Gorge where I live provides easy access to conifer forests, with low-elevation rain forest in the western gorge to semi-desert in the eastern gorge, plus mountain habitat to the north and south. The incredible diversity of this region produces a correspondingly incredible diversity of fungi.

Each season begins with early morels, *Verpa bohemica*, hiding under cottonwood trees when their young leaves reach the size of mouse ears (in March near the Columbia River and as late as June in the mountains). Hunting for true morels starts near the river in April with the appearance of the landscape morel, *Morchella importuna*, around town and in low-elevation areas logged the previous year. When trillium blossoms start to turn pink and the *Calypso bulbosa* orchid is blooming, four of my favorite morel species are in peak production. The trillium blooming stage also tells me when the burn morels will start to appear at a given elevation. (They grow one to two years after a forest fire.) One delicious burn morel even grows from summer into fall, sometimes until snowfall, and starts to fruit only after the trilliums have faded away.

In April, spring king boletes begin to appear in the hills immediately north and south of the Columbia River, followed by fruiting at around 2,000 feet in May and 3,000 feet in June. I collect and eat more spring king boletes than any other mushroom. As soon as the spring king season is over, it is time to pursue king boletes, starting at around 4,000 feet in mid-July and August and then moving down to about 3,000 feet by October, with queen boletes appearing at low elevations in October and sometimes hanging on through December.

Chanterelles are abundant from August until hard winter freezes. Meanwhile, starting in August, I seek bear's head mushrooms in the mountains, followed by low-elevation lion's mane under the oaks from October through December. One mild January I found nice large clumps of oyster mushrooms growing on dead and dying cottonwood trees along the river, only to be starting again fresh by April.

Edibles are far from the only fungi that catch my eye. I seek out dye mushrooms for my artist friends. A wide range of different fungi produce interesting colors for dying silk and wool. Some of the most famous are *Phaeolus schweinitzii* (the dyer's polypore) and *Cortinarius* species in subgenus *Dermocybe*. I harvest medicinal mushrooms for personal use, notably any species in the genus *Hericium* and any member of the *Boletus edulis* group (conveniently, *Hericium* and *Boletus* species are also among my very favorite edibles).

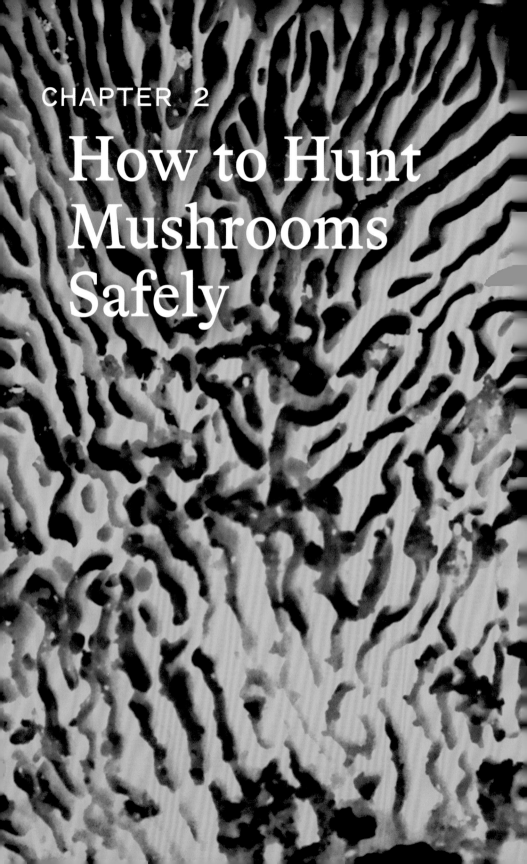

CHAPTER 2

How to Hunt Mushrooms Safely

For those of you who are wondering about how many people on average die each year from mushroom poisoning in the United States and Canada, the answer is one or two people. Fortunately, we have good books and other resources to guide us, unlike in much of the underdeveloped world.

Consider joining a mushroom club near where you live so you can learn from experienced foragers and get help confirming identifications. The North American Mycological Association, or NAMA (namyco.org), has lists of clubs by city, state, and province—and it's also a good organization to consider joining (https://namyco.org/clubs.php).

Good targets for new foragers include all of the abundant and distinctive mushrooms that I personally gather for the table: chanterelles (any *Cantharellus* or *Craterellus* species); true morels (any *Morchella* species); spring king boletes, king boletes, and queen boletes; and any *Hericium* species. Oyster mushrooms are distinctive, abundant, nutritious, and medicinal. Almond-scented *Agaricus* species are prized as both food and medicine—but, unfortunately, I'm allergic to them. And the shaggy mane, *Coprinus comatus*, is great for oyster lovers.

On the other end of the scale, you should also learn to identify destroying angels and death caps, the deadly *Amanita* species, right away. Never eat *any* amanita until you are **very** experienced at identification. And do not eat any little brown mushrooms. Many little brown mushrooms, especially those in the genus

HELPFUL RESOURCES

To view any of the more than twenty free YouTube videos I've made on introductory to advanced topics, go to the NAMA website (https://namyco .org/educational_programs.php).

If you are planning to forage mushrooms for food, visit the NAMA website to read the information I wrote about all the known seriously poisonous fungal species of North America (https://namyco.org/mushroom_ poisoning_syndromes.php).

I've also posted educational videos and blogs, my scheduled presentations, and contact information on my website, www.mushroomsof cascadia.com. Be sure to view "Should I Have Eaten That?" (https://www. mushroomsofcascadia.com/about-the-author/videos-and-interviews). (You can also find this video by Googling "Should I Have Eaten That?" plus "Beug.") The presentation covers incredible edibles and mistakes to avoid (three versions are available).

Inocybe, can make you very sick (and can even kill dogs). Funeral bells (*Galerina* species in subgenus *Naucoriopsis*) could kill you, as one insufficiently observant hallucinogenic mushroom–seeker discovered. Deadly *Amanita* and *Galerina* species, or close look-alikes with the same properties, are found in all north temperate and arctic regions. *Inocybe* species are abundant globally.

Preparing for the Hunt

Before you head off into the forest or fields to hunt for mushrooms, you'll want to gather a few helpful items. For transporting your finds, I recommend a broad-bottomed container such as a five-gallon bucket (with drain holes drilled in the bottom) or a basket. Add a knife and a brush to your foraging kit (you can make a great knife-brush combination by taping a half-inch paintbrush to the end of an old steak knife) for harvesting and cleaning. Throw in a powerful whistle (or even an air horn) for safety, plus a compass and some small paper bags (waxed paper sandwich bags work well). It's also useful to bring one or two larger bags for edibles or for larger specimens (I carry two lightweight cloth bags). Alternatively, a roll of aluminum foil works well too.

The idea is to keep each species separate and protected from damage in your basket. Plastic bags are problematic because mushrooms will sweat and decompose quickly in them. Fungal fruitbodies are still alive and respiring even after harvest. Ideally, carry a small notebook to record the date, location, and habitat where you harvested each mushroom. The habitat is often useful for differentiating one species from another. And including a location and date will help you remember when and where that species might be found in the future. Consider bringing a camera (ideally equipped with a macro lens or close-up adapter) so that you can make a photographic record of your finds. And, finally, bring a map of the area where you plan to hunt, and study it carefully before heading into the woods.

YOUR FORAGING CHECKLIST

- ☐ Basket or five-gallon bucket with drain holes
- ☐ Knife
- ☐ Brush
- ☐ Whistle or air horn
- ☐ Compass

- ☐ Assorted small and large paper or cloth bags, or aluminum foil
- ☐ Notebook and pen
- ☐ Camera
- ☐ Map of the area

Staying Safe

It is not unusual for mushrooms hunters to get lost, so as an extra measure, I use a GPS app on my smartphone. I preload maps of the areas where I plan to hunt and then place a waypoint before heading into the woods. Once in the woods, the GPS tracks my route and points out the way back to my car. Be aware, however, that your smartphone batteries may die or you could lose satellite coverage, so always bring a compass and back-up plan. I like to do my hunting within sight of established trails or earshot of traveled roads. Go with a companion and stick within earshot of each other, checking in regularly if you are collecting apart from your companion. Use whistle signals: one blast means "Where are you?", two blasts says "I am here!", and three blasts is the international call for help. Be aware that in the woods even a strong whistle blast does not carry a great distance.

Making Your Collection

When you find familiar edible mushrooms that you are planning to harvest and eat, simply cut off the mushroom at ground level with your knife. Carefully brush off all dirt and debris and place it in your edibles bag. Collecting nice, clean material will make preparing the mushrooms for the table much easier. One exception to this rule: When I am collecting edible boletes or matsutake, I use my knife blade to pry the entire mushroom from the ground and immediately trim off the dirt. I eat every bit of these mushrooms. Some foragers use a narrow rake to unearth truffles, young boletes, or matsutake. Doing so will destroy some of the mycelial network (the mushroom organism itself) and thus greatly decrease future harvests. Also, raking for truffles (rather than using a trained dog) yields mostly tasteless, valueless immature truffles.

For mushrooms you collect for study, use your knife blade and carefully pry out the entire mushroom, being careful not to rub off any veils or loose material on the stipe (the stem) or the cap. For a mushroom with white gills that you suspect may be an *Amanita* species, be certain to collect any underground remains of the volva (the cup-like structure at the bottom of the mushroom). Also be aware that, rarely, the mushroom will have a long underground portion or will be growing from a sclerotium (a buried, dense, tuber-like mass of mycelial tissue). Try to collect these underground portions if present. Also try to collect at least three specimens of each species, ideally representing young, maturing, and mature material.

If the mushroom is one that you want to spore print, to determine the spore color, tear off a small sheet of white paper, place the paper under the cap of the mushroom, wrap both paper and mushroom inside a paper bag, and place it in a container with the gills or other spore-bearing surface pointing down. Often your mushroom will have deposited enough spores to determine their color by the time you are home. Even white spores will show clearly on white paper (and to be certain, can be rubbed off, leaving a white deposit on your finger). And many times you will see a spore deposit (a mature mushroom produces millions to trillions of spores) simply by looking at the vegetation immediately under the mushroom you are about to harvest, at a mushroom cap overlapped by your mushroom, or at the mushroom stipe, eliminating the need to make a spore print. Also, when you are confident that you have both a young and an old specimen of the same species, any change in gill color usually tells you the spore color.

Recording Your Specimens

As you open each bag of mushrooms back home, smell and record the odor of each type of mushroom (in the woods it is often too cold to determine the odor accurately). Common odors are earthy, woodsy, or mild. Odors can be very distinctive of anise, almond extract, cucumber, creosote, baby vomit, and so on. Then taste the mushroom by taking a small nibble from the cap. Chew and spit out the mushroom fragments. While tasting (but not ingesting) is safe to do with even poisonous mushrooms, you risk harm to your health and even death if you swallow poisonous mushrooms.

Record the taste. The taste is often mild or similar to store-bought mushrooms, or it can be bitter, peppery hot, or unpleasant like rancid flour (farinaceous). Note that if the odor is repulsive, the taste will also be similar. Tasting is safe to do even with deadly poisonous mushrooms. Caution: Do not taste large, red-pored boletes (the reaction can be most unpleasant). Also, if you believe that your mushroom may be a species of *Russula* or a *Lactarius* (both of which are unusually brittle with flesh that breaks cleanly, if unevenly, into two or more disconnected pieces, like breaking a segment of crisp apple or a piece of chalk), taste the stipe first and taste the cap only if the stipe is not peppery hot, since tasting the cap of exceptionally hot species like these can blister your tongue. WARNING: DO NOT TASTE-TEST UNKNOWN WILD PLANTS— YOU COULD DIE BY TASTING SOME POISONOUS PLANTS, BUT NOT FROM TASTING MUSHROOMS.

Next, get edibles into a refrigerator to keep them fresh until it's time for cleaning and/or cooking. For unknown species, get a spore print started if you have not already determined the spore color in the field. Place the mushroom on a piece of white paper with the gills or other spore-bearing surface pointing down. Cover the mushroom with a cup or bowl to keep it from drying. Most mushrooms will yield a visible spore print in an hour or two. Immature mushrooms can take much longer (some *Gyromitra* species take a week to mature enough to drop spores).

Identifying Your Find

Start with the picture key in this book and see if you can identify a mushroom while you're still in the field or once you are home. Many of the mushrooms in this book can be positively identified just from the photos and the descriptions in the key without requiring you to consult any external resources. But my hope is that you will use this book in conjunction with the free, downloadable mushroom program MycoMatch, a product of the Pacific Northwest Key Council, of which I am one of the authors. MycoMatch has more detailed descriptions than you will find in any mushroom field guide or reference guide. It also contains extensive references to the mycological literature. MycoMatch has descriptions of more than 4,200 species found in the Pacific Northwest, with more than 6,800 photographs of 2,500-plus of the included species. In cases of well-known mushrooms, it offers multiple images of each species, showing the variations in form and color. Also, for every entry, creator Ian Gibson has listed whether the species is present in British Columbia, Canada (BC), Washington State (WA), Oregon (OR), and/or Idaho (ID) plus other states, provinces, and countries where a given species (or a close look-alike) has been documented in his references. This makes both this book and MycoMatch useful in all north temperate regions. The picture key also works well as a supplement to any other mushroom reference guide.

When collecting for science or to broaden your knowledge of mushrooms, record the dimensions of the mushroom as well as the colors of the cap surface, cap flesh, stipe surface, and stipe flesh. Consider examining the specimen with a ultraviolet (UV) light or UV flashlight (such as a scorpion-hunting light) in a darkened room. This can not only be beautiful but is also a powerful identification aid.

Chemical tests and microscopy can be essential for a positive identification of less distinctive species and for uncovering look-alikes. The common and most versatile tests are listed opposite. Do any chemical tests on small pieces of mushroom that you cut off (and not on a mushroom that you want to dry and save or submit for DNA analysis). For accurate results, always measure spore size and shape from fresh spore prints to make certain that you are examining mature spores. I like to make my spore prints at least partially on a glass slide to make measuring the spores easy.

Useful Chemical Tests for Scientific Study

Use of a few common chemicals can help with identification, but this is not needed for most mushrooms (so long as you stick to the many distinctive species and are not doing any microscopy). To study ascomycetes under a microscope, always use a few drops of water or Lugol's solution (0.25 percent iodine and 0.5 percent potassium iodide in distilled water) on your glass slide with your specimen. Lugol's can also be useful for basidiomycetes if you cannot obtain Melzer's solution. For studying basidiomycetes, you can use 3 percent potassium hydroxide (KOH), water, or Melzer's solution. Melzer's is just Lugol's solution plus some chloral hydrate. Free starch (and starch-like substances in some mushrooms) causes a blue reaction with either Lugol's or Melzer's. Other chemicals in mushrooms can cause a red reaction with these reagents, and these color changes often help you tell species apart. I sometimes use a drop or two of 10 percent KOH or 10 percent iron sulfate dripped on a piece of the mushroom to look for a color change (you can substitute household ammonia for KOH and buy iron sulfate from a fertilizer store). Melzer's can be hard to obtain, but leaders in most mushroom clubs can direct you to a source for Melzer's or any of these solutions.

CHAPTER 3

Mushrooms in the Kitchen

There are a few ways to preserve any mushrooms you don't immediately plan to use, including drying and freezing. I rinse off mushrooms I am planning to cook under cold running water and then brush them off with a foam brush, a soft-bristle brush, or a moistened paper towel. Morels, chanterelles, boletes, and other prized edibles can be stored for up to a week or more in a paper bag in the refrigerator (if they are worm-free). Discard any mushrooms showing signs of spoilage. Exception: Shaggy manes decompose quickly and must be eaten right away or stored immersed in water.

Drying and Freezing

To prepare any mushroom that is to be dried, I skip the cold-water rinse and clean them with a moistened, soft-bristle brush or paper towel. For edibles, I do a final trim with a paring knife to remove any lingering dirt and cut out any bad spots, though I skip the trim for mushrooms being dried for scientific purposes. If I am cleaning boletes, I cut the cap from the stipe and look for holes from insect larvae. If I find any wormy specimens, I promptly cook or dry them before the insect larvae consume too much of my mushroom. (If you are not squeamish, the fried larvae are nutritious and taste just like the mushroom they are living in. If you are squeamish, slice and dry the mushrooms, then shake out and discard the dried larvae.) Hint: I slice the bolete caps and stipes about ⅜ inch thick and then often sauté the fresh bolete stipes, which are firm and high in healthful beta-D-glucans, and dry the caps. For other edible mushrooms, I slice the caps and stipes to about ⅜ inch thick and dry them until brittle.

I never dry chanterelles because they do not reconstitute well after drying (though you can grind up brick-hard dried chanterelles to use as a mushroom seasoning).

I usually dry morels whole, stems intact. If you want to make certain that there are no occupants inside a hollow morel, cut it in half from top to bottom. Halving is also useful when dealing with big, fat morels.

Mushrooms are best dried in a food dehydrator set at up to 105°F for twelve to twenty-four hours. Dry until crisp and brittle. Note that you can use higher drying temperatures if you are simply drying mushrooms for food and not for later scientific study. After drying, I seal the specimens in resealable plastic bags (once mushrooms are dried, plastic bags are okay) and place the bags in the freezer for a few days. Freezing at this stage will kill any insect eggs or larvae that survived the drying process. I keep my dried mushrooms in the freezer until I'm ready to use them, but you can also store them in glass jars after the initial freeze.

The flavor of dried boletes gets better over time, and they can be stored in a glass jar at room temperature for a few years. On the other hand, morels can become too strong to be pleasant if stored at room temperature, so I keep my dried morels in the freezer.

USING TRUFFLES

Truffles are valued purely for their aroma, and when they are immature, they have no odor or flavor. Ripe truffles, on the other hand, have a cheesy, garlicky, sulfurous odor that comes partially from a compound in the truffles that is the same as the male sex pheromone of pigs. It is a "love it" or "hate it" aroma—and I'm in the former camp. I like to nest fresh Oregon white truffles in a dry paper towel and place it in a large container along with butter, cheese, whole raw eggs, or whole ripe avocados—foods with high oil content. I seal the container and refrigerate it for two days to enable the ingredients to take on the aroma of the truffles.

I can usually repeat this process two or three times before the truffle odor starts to diminish. Then I start using the truffles themselves. A simple breakfast favorite is scrambled truffled eggs. Grate a small amount of truffle onto the eggs after cooking and enjoy them with toast with truffled butter. At dinnertime, I like to cook a batch of noodles, butter them with truffled butter, and grate truffled cheese (and truffles) on top. Asiago cheese is one of my favorite cheeses to truffle. Hard cheeses accept the truffle aroma better than soft cheeses. And you can store the remaining truffles and truffled butter and cheese in the freezer for months.

Truffling foods (truffles nested in a paper towel)

For long-term storage of surplus mushrooms, dry or cook them and then store them in a resealable plastic container in the freezer. Freezing uncooked mushrooms is dangerous, because the enzymes are still active in raw mushrooms and spoilage can occur.

Reflections on Canning and Pickling Mushrooms

I don't recommend canning mushrooms. When foods are canned, the air is removed from the jar, which provides a suitable medium for growth of mycotoxin-producing molds, *Salmonella* bacteria, or the bacterium *Clostridium botulinum*. Dormant spores are everywhere, covering all the foods we eat. Because the spores germinate only in the absence of air, fresh foods pose no problems. Sticking your finger into a jar that you think may have been contaminated and then licking your finger could be enough to kill you if you did not first eliminate the *C. botulinum* spores or prevent their growth by keeping the canned goods sufficiently acid. These spores can be killed by heating the canned foods at 240°F for up to two hours (a process that requires a pressure canner). So unless you have the right equipment and are a very experienced canner, I'd choose other means of preservation.

Pickling and then canning mushrooms in a hot water bath is one way to store them safely in an acidic environment. Even if *Clostridium* spores are not killed off during the hot water bath, keeping the pH at less than 4.6 via pickling can prevent spore growth. Use a pickling liquid comprising 50–60 percent vinegar (5 percent acidity) in distilled water plus adequate added salt to prevent the vinegar from being metabolized (¾ tablespoon of salt per cup of pickling solution).

In addition to making pickles using vinegar, they could also be made via fermentation. The fermentation that occurs during pickle making utilizes bacteria of the Lactobacillaceae family to produce enzymes that oxidize alcohols to lactic and acetic acid. The bacteria are stimulated by the addition of salt. Thus, salt serves two roles in a pickle recipe—it draws water from the mushrooms or other foods by osmosis, and it triggers fermentation. The fermentation in turn produces acids that help preserve the food. Fermentation also generates complex flavors and releases vitamins.

If you decide to try making mushroom pickles, make certain that every jar is properly sealed. Prior to use, inspect the jar to be sure the lid is still sealed (lid button down) and check for mold growth (visible growth on the surface of the canned product) or yeast growth (cloudy liquid) and off-odors. Discard any suspect containers without tasting what's inside! Pickling is safer than canning but great caution is still needed.

Cooking Mushrooms

Except for mushrooms prized for their aroma, such as matsutake and truffles, my favorite method of cooking is to slice the mushrooms about ⅜ inch thick and place them in a dry skillet that's large enough to accommodate the slices in a single or double layer.

Dry sauté (don't add anything to the skillet) the mushrooms over medium heat for a few minutes to release and evaporate most of the water, and then add butter and/or olive oil (just enough to thinly coat the bottom of the pan) and cook until the mushrooms are golden brown. The idea is that, at the browning stage, the mushrooms will completely cover the bottom of the pan so that each slice can be nicely browned on both sides. (Alternatively, cut the mushrooms into ⅜-inch cubes and stir them around to brown each side; you do not have to turn each piece carefully.)

Salt the sautéed mushrooms lightly and serve as is, add them to scrambled eggs or an omelet, or place them on your favorite toast, sprinkle hard cheese on top, and melt the cheese under the broiler. These methods enable the delicate mushroom flavors to be fully experienced.

Following are a few more of my favorite mushroom recipes.

Stuffed Morels

SERVES 6 AS A SIDE DISH

18 medium or 12 large fresh morels (reconstituted dried morels will also work)

¼ cup finely crushed crackers (I like garlic and herb crackers)

4 eggs

¼ cup finely minced pecans or crab meat

¼ cup finely minced green onion

¼ cup finely grated cheese (use a few of your favorite varieties)

1 teaspoon sea salt

A pinch of pepper, whatever kind you like

1 tablespoon butter, olive oil, or duck fat for sautéing

Rinse the mushrooms under cold running water and brush off any dirt with a soft-bristle brush. Pat dry with a paper towel. Cut off the stems and set the mushrooms aside. Mince the stems finely.

Fill a shallow bowl with the crushed crackers (sometimes I sift them through a meshed metal strainer to remove the big chunks).

Crack 4 eggs into a separate bowl and beat until combined.

In a mixing bowl, combine the minced morel stems, the pecans (or crab meat), onions, cheese, salt, and pepper. Carefully stuff each morel with the mixture, using a wooden skewer or chopstick for those with long or narrow openings. Pack them well but take care not to split them open. Note: If necessary, split one side of each morel with scissors to facilitate stuffing.

In an iron skillet over medium heat, warm enough butter to thinly cover the bottom of the pan.

Roll the stuffed morels in the eggs until coated, and then drop them into the bowl of crushed crackers and roll them around until thoroughly coated. Place the coated, stuffed morels in the skillet in a single layer and cook until they are golden brown on all sides, about 15 minutes.

Serve immediately.

Mushroom Duxelles

MAKES 6 SERVINGS PER POUND OF MUSHROOMS

1 to 5 pounds morels, chanterelles, king boletes, or any of your favorite edible mushrooms

⅓ to 2 pounds sweet onions or shallots

Olive oil, avocado oil, and/or butter for sautéing

Good-quality dry white wine or sake

Salt

Freshly ground nutmeg (optional)

Select one skillet or a pair of skillets that are just large enough to hold all of the mushrooms, with the skillet(s) no more than three-quarters full.

Rinse the mushrooms under cold running water and use a soft-bristle brush to remove any dirt. No need to towel-dry the mushrooms after rinsing. Finely chop the mushrooms (or pulse them quickly in a blender or food processor). Measure the mushrooms, and then finely chop about one-third as much sweet onion or shallots.

Set the skillet(s) over medium heat and add the mushrooms and onions. Cook, stirring, until the liquid has cooked off and the mushrooms are beginning to stick to the skillet (10 to 20 minutes). Add a little olive oil and stir the mushrooms, but do not let them brown.

Pour in enough wine to cover the mushroom mixture. While lubricating the cook and your guests liberally with the remaining wine, continue cooking until the wine has reduced and the mushrooms are beginning to stick and brown. Season to taste with salt and nutmeg (I use about 1 teaspoon nutmeg in a 10-inch skillet) and remove from the heat.

You now have duxelles ready for use on toast with melted extra-sharp cheddar or in a quiche, omelet, or egg scramble.

To store, freeze the duxelles in ice cube trays and then transfer the cubes to a closed container and store in the freezer for up to a year.

Mushroom Quiche

This recipe was inspired by Ruth Bass's book, Mushrooms Love Herbs.

MAKES ONE 10-INCH QUICHE; SERVES 6-8

1.5 ounces dried morels or king boletes, ¾ pound fresh mushrooms (chanterelles work well), or ¼ pound Mushroom Duxelles (see recipe, opposite)

2 tablespoons butter

1 unbaked 10-inch pie shell

2½ cups shredded Swiss, asiago, or provolone cheese, or a mixture

4 eggs

1 cup light cream or quark

¾ cup milk

1 tablespoon minced fresh sage

2 tablespoons fresh chives or 1 tablespoon minced parsley

¼ teaspoon salt

¼ teaspoon freshly ground black pepper

¼ teaspoon dry mustard

Preheat the oven to 350°F.

If using dried mushrooms, reconstitute them in a bowl with 1 cup water. If using fresh mushrooms, rinse them under cold running water and use a soft-bristle brush to remove any dirt. Roughly chop the mushrooms.

In a skillet, melt the butter over medium heat. Add the mushrooms and sauté for about 15 minutes, until the mushrooms are golden but not browned. (If using freshly made duxelles, you can skip this step. If using frozen and then thawed duxelles, sauté just enough to lightly brown the duxelles; about 5 minutes.)

Cover the bottom of the pie shell with the shredded cheese and the mushrooms.

In a bowl, combine the eggs, cream, milk, sage, chives, salt, pepper, and dry mustard and beat until smooth. Pour the mixture into the pie shell and bake the quiche for 40 to 50 minutes, until firm and browned.

Serve warm.

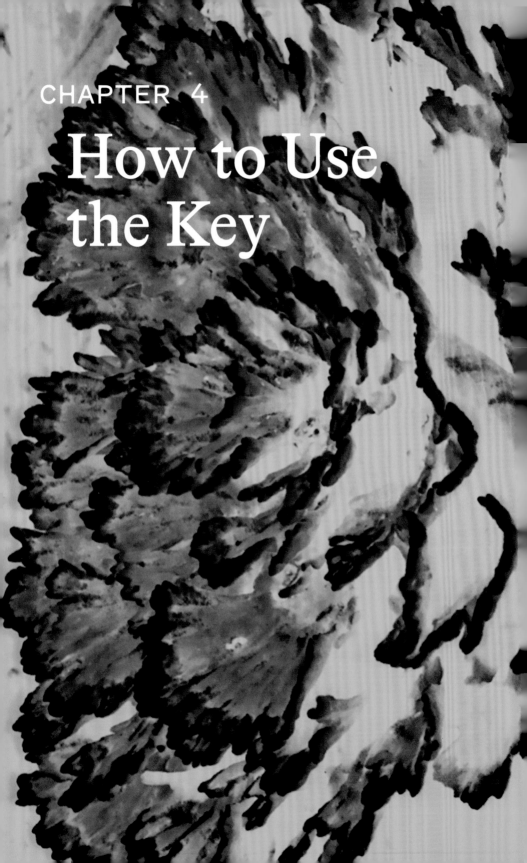

CHAPTER 4

How to Use
the Key

In this unique identification tool, carefully cropped photos illustrate the features that distinguish one mushroom species from another, while the text in the key provides information not obvious from just a photo. A series of choices will efficiently guide you to the best mushroom identification.

To pack information into as few words as possible (and thus have room for a lot of species in the book), I employ some shortcut symbols:

±	means either "with or without" or "more or less"
≤	means "less than or equal to" or "up to"
≥	means "greater than or equal to"
<	means "less than"
>	means "greater than"
=	means "previous name, a synonym"
/	separates two look-alike species distinguished by DNA
ID	means "identification"
GI	means "gastrointestinal distress"
EDIBILITY?	means "Edibility unknown or untested"
GENUS SPECIES?	means that the correct name is not firmly established
NA	means "North America"
EU	means "Europe"

To see how the key works, let me guide you through two examples.

Mystery mushroom X

Mystery Mushroom X

Note: A spore print will prove to be white for mushroom X. If you cut mushroom X in half from top to bottom, you will see that the gills are free from the stipe, but just barely. Your mushroom was found under oaks and was growing in October.

Turn to the first page of the key (page 42). On this page, you'll see this series of photos and descriptions.

1a. Fleshy, with thin blade-like gills on the underside of the cap (page 166) *175a*

1b. Non-gilled fungi *2a*

Key lead 1a, the very first listing, describes fleshy mushrooms with thin, blade-like gills on the underside of the cap. Does mushroom X have gills? By reviewing the pictures shown, you see similar-looking mushrooms in key lead 1a. So, yes, your mystery mushroom has gills. In the right margin of key lead 1a, you see (page 166) 175a, which tells you to turn to page 166 and start reading at key lead 175a, dealing with spore color. (Incidentally, if at any time you don't understand terms used in these descriptions, simply turn to the glossary on page 322 for definitions.)

Here is the part of page 166 with key leads 175a through 178b.

175a. (1a) Spores dark chocolate-brown to black (page 293) *385a*

↓175b. Spores light brown, orange-brown, or rusty brown (page 257) *327a*

↓175c. Spores pinkish salmon (page 251) *315a*

↓175d. Spores white or olive-green; gills free from stipe (page 166) *177a*

175e. Spores white, cream, pale yellow, pale orange, or pinkish buff; gills attached to stipe (with a notch, bluntly or decurrent) *176a*

176a. (175e) Growing on wood (or from buried wood) (page 245) *305a*

↓176b. Breaks crisply (*Russula* and *Lactarius* spp.) (page 229) *276a*

↓176c. Gills thick, often brightly colored; sometimes viscid cap; fleshy stipe (*Hygrophorus* and *Hygrocybe* spp. and relatives) (page 219) *262a*

↓176d. Caps usually < 1 in. wide, often conical; fragile cap and stipe; no partial veil (Mycenoid) (page 213) *253a*

↓176e. Caps < 1 in. wide, may be depressed to umbilicate; sometimes decurrent gills; no veil; thin stipe (Omphalinoid) (page 207) *240a*

↓176f. Caps small to large, convex to flat; gills generally bluntly attached and usually closely spaced (Collybioid) (page 200) *223a*

↓176g. Caps > 1 in. wide; may have a partial veil, sometimes viscid; gills usually notched; fleshy stipe (Tricholomatoid) (page 190) *207a*

176h. Caps > 1 in. wide, like Omphalinoid; adnate to decurrent gills, rarely viscid; rarely veiled (Clitocyboid) (page 180) *196a*

178a. (177b) Volva not a membranous cup (page 169) *185a*

178b. Volva a membranous cup (page 167) *179a*

Read the five choices (175a through 175e) and pick the best fit. You know from the previous note that mushroom X has white spores, and cutting the mushroom in half top to bottom showed that the gills are barely free from the stipe. So, in this case, the best fit is 175d, which then directs you to lead 177a, on the same page.

177a. (175d) Gills distinctly free; cap breaks from stipe in a ball-and-socket fashion; sometimes scaly, not viscid (*Lepiota* spp. and allies) (page 174) *190a*

↓177b. Gills barely free of stipe; cap does not separate from stipe in ball-and-socket fashion; membranous universal veil present (*Amanita* spp.) *178a*

177c. Gills barely free of stipe; viscid universal veil (page 179) *195a*

If 177a doesn't describe your mystery mushroom, keep reading down through the choices (177b and 177c) and pick the best fit. Note that in lead 177a, the lead in parentheses (175d) tells you where you just came from in case you need to back up. Because the gills in mushroom X are barely free of the stipe, the best fit is 177b, which tells you that you have a mushroom of the genus *Amanita* and directs you to key lead 178a.

178a. (177b) Volva not a membranous cup (page 169) *185a*
178b. Volva a membranous cup (page 167) *179a*

The volva in the photo is a membranous cup, so you proceed to key lead 179a on the following page.

179a. (178a) Partial veil thin or absent; prominent cap striations (page 168) *182a*

179b. Thick and membranous partial veil (and universal veil) *180a*

Since both the partial veil and universal veil are thick, you go to key lead 180a.

180a. (179b) Fruits late summer to winter under oaks and other hardwoods

Amanita phalloides

Accidental introduction from EU, found on the NA East Coast and West Coast, from Los Angeles to Vancouver, BC, as well as many other parts of the world. Cap can be white, greenish, or bronze; ± volval patch on cap. DEADLY!

180b. Fruits late winter to spring in mixed woods, usually with oaks *181a*

Since you collected your mushroom in October, you conclude (correctly) that you have found a deadly amanita and, from the comments, learn that it can be white, greenish, or bronze. Your other choice (180b) directs you to key leads 181a and 181b. The mushroom in 181a is also deadly (and has a deadly look-alike), while the mushroom in 181b is edible (and has an edible look-alike). Can you tell which *Amanita* species shown below is edible and which is deadly?

A. phalloides　　　　*A. calyptroderma*

The image on the left is of *A. phalloides*, the species that accounts for 90 percent of all mushroom deaths. The mushroom on the right, *A. calyptroderma*, is edible. Eating *Amanita* species is only for experts! Consult MycoMatch to help you confirm that you have made correct identifications.

　Mystery mushroom Y is clearly also a gilled mushroom. It is growing on wood (a very decayed hemlock log). If you cut one off flush from the wood, you would see that there is just a plug of a stipe, the gills are attached (not free), and the edges are

Mystery mushroom Y

smooth, not saw-toothed. The odor is pleasant, and the taste is interesting but very hard to describe.

Turning again to page 42 and the start of the key, you see that for all gilled mushrooms you turn to page 166 and read entries 175a to 175e and pick the best fit. Because mushroom Y's gills are attached to the stipe, your best choice is 175e, which directs you to key lead 176a.

176a. (175e) Growing on wood (or from buried wood) (page 245) *305a*

↓**176b.** Breaks crisply (*Russula* and *Lactarius* spp.) (page 229) *276a*

↓**176c.** Gills thick, often brightly colored; sometimes viscid cap; fleshy stipe (*Hygrophorus* and *Hygrocybe* spp. and relatives) (page 219) *262a*

↓**176d.** Caps usually < 1 in. wide, often conical; fragile cap and stipe; no partial veil (Mycenoid) (page 213) *253a*

↓**176e.** Caps < 1 in. wide, may be depressed to umbilicate; sometimes decurrent gills; no veil; thin stipe (Omphalinoid) (page 207) *240a*

↓**176f.** Caps small to large, convex to flat; gills generally bluntly attached and usually closely spaced (Collybioid) (page 200) *223a*

↓**176g.** Caps > 1 in. wide; may have a partial veil, sometimes viscid; gills usually notched; fleshy stipe (Tricholomatoid) (page 190) *207a*

176h. Caps > 1 in. wide, like Omphalinoid; adnate to decurrent gills, rarely viscid; rarely veiled (Clitocyboid) (page 180) *196a*

Of choices 176a to 176h, only key lead 176a includes species growing on wood (or from buried wood), so you turn to page 245 and key lead 305a.

305a. (176a) Species bluntly attached to wood or with a stubby stipe (page 248) *309a*

↓**305b.** Species with a distinct stipe; saw-toothed gill edges (page 247) *308a*

305c. Species with a distinct stipe; smooth gill edges *306a*

Of choices 305a to 305c, only 305a fits (bluntly attached to wood with a stubby stipe), so you proceed to page 248 key lead 309a.

309a. (305a) Stipes rudimentary to absent; gills with smooth edges *310a*

↓309b. Gills with saw-toothed edges; cap (1–4 in. wide) brown; often near snowbanks

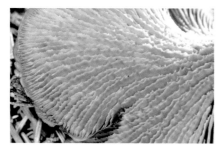 

Lentinellus vulpinus   *Lentinellus montanus*

L. vulpinus (EU, NA) is on hardwoods and *L. montanus* (NA) on conifers. In both, cap centers are coarsely hairy, odor slightly aromatic, taste slowly peppery. Thin, tough, inedible.

309c. Gills split in two lengthwise; cap (< 1½ in. wide) hairy, whitish; on hardwood

Schizophyllum commune group

The caps are fan-shaped, thin, tough, and leathery. The gills, when young, are whitish to grayish. The odor and taste are pleasant to sour. Too small and tough to be of interest. Several closely related, similar species share this name. Worldwide distribution.

310a. (309a) Gills white or very pale color (page 249) *312a*

↓310b. Gills yellow to orange (page 249) *311a*

310c. Gills silvery or pale tan, darkening in age; cap (< ¼ in. wide) dark

Hohenbuehelia unguicularis (± 2x life-size)

Fruits spring to winter on hardwoods in northern NA and EU, but rare or rarely noticed due to tiny size and dark color. Distinctive. Odor and taste not noted. Too tiny to eat. Very similar *Resupinatus applicatus* group members (global distribution) are distinguished microscopically.

Since the gills are white, you proceed to (page 249) 312a.

312a. (310a) Off-white, pinkish tan, or tan (page 250) *314a*

312b. Pure white *313a*

Since you have a pure white species, you proceed to 313a.

313a. (312b) Minute hanging or upright pitcher-shaped; lacking gills

Calyptella capula (± life-size)

Grows on wood or dead plant material in damp areas, summer to fall in NA, EU. Related to *Marasmius* species. Odor and taste not noteworthy. Easily mistaken for an ascomycete.

↓313b. Cap (< ¾ in. wide) dry, minutely hairy, not peelable; short lateral stipe

Cheimonophyllum candidissimum (± life-size)

Gills distant, yellowish in age. Fruits July to October on hardwoods and conifers in Cascadia. Indistinct odor and taste. *Panellus mitis* (Cascadia, EU) looks very similar but has a peelable cap due to a rubbery, gelatinous layer. Too tiny to eat.

313c. Cap (1–4 in. wide) smooth, may turn creamy in age

Pleurocybella porrigens PNW01

Angel wings, normally on hemlock logs (NA, EU, Asia), have a pleasant odor and unusual flavor. Long a popular edible. In Japan one year, however, several elderly people on dialysis who ate a huge quantity slowly died from holes in their brains. The species in Cascadia is genetically distinct.

Of choices 313a to 313c, the latter, *Pleurocybella porrigens*, is the best choice, and you confirm the identity of mushroom Y by looking at the full description in MycoMatch. This is an edible species, but no mushroom should be eaten in huge quantities, especially by the elderly. You'll read about a unique Japanese case that involved elderly people who consumed huge quantities.

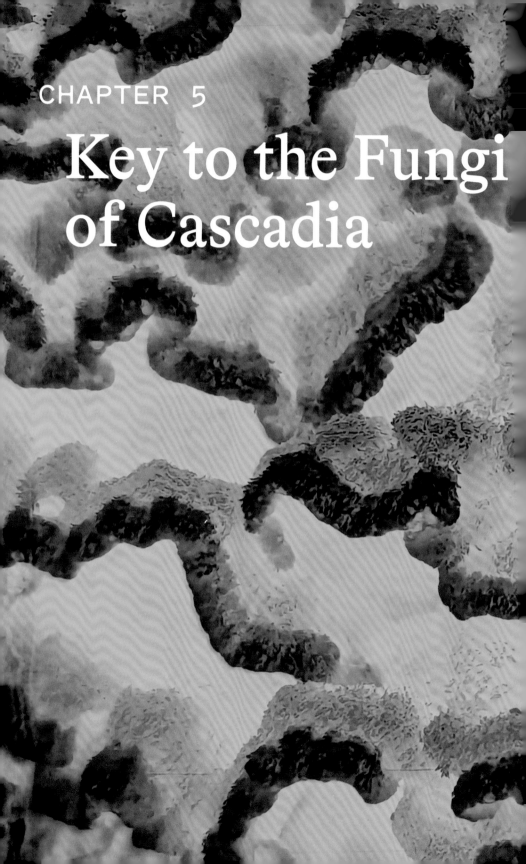

CHAPTER 5

Key to the Fungi of Cascadia

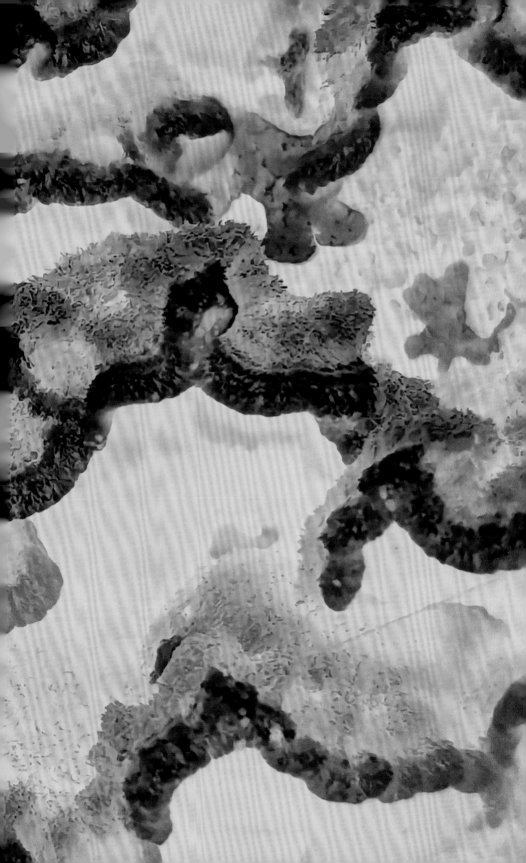

1a. Fleshy, with thin blade-like gills on the underside of the cap (page 166) *175a*

  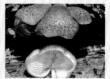 

1b. Non-gilled fungi *2a*

2a. Fruitbody an unbranched fleshy to woody club (page 162) *168a*

↓2b. Fruitbody gelatinous (the jelly fungi) (page 156) *154a*

↓2c. Fruitbody with spines under cap, or hanging icicles (page 146) *140a*

↓2d. Fruitbody coral-like with upward-pointing tips or lobes (page 132) *118a*

Illustrated species for key entries 1a through 3e (listed left to right)

1a. *Amanita muscaria, Cortinarius albofragrans, Crepidotus calolepis, Laccaria laccata*
2a. *Clavaria rosea, Clavaria fragilis, Clavariadelphus truncatus, Xylaria hypoxylon*
2b. *Ascocoryne sarcoides, Auricularia americana, Dacrymyces aquaticus, Dacryopinax spathularia* 2c. *Echinodontium tinctorium, Hericium erinaceus, Hydnum washingtonianum, Hydnellum caeruleum* 2d. *Clavulina coralloides, Ramaria araiospora, Sparassis radicata, Ramaria purpurissima*

↓2e. Fruitbody a tiny bird's nest; ± round, inside a solid to powdery mass (puffballs, earthballs); or stinking; erupting from a sack-like volva (page 124) *106a*

↓2f. Fruitbody ± buried, ± globular; or a woody desert species (page 118) *97a*

↓2g. Fruitbody woody or leathery, or soft with inseparable pores (page 94) *66a*

2h. None of the above, fruitbody fleshy (page 44) *3a*

2e. *Nidula niveotomentosa, Calbovista subsculpta, Mutinus caninus, Scleroderma cepa*
2f. *Geopora cooperi, Leucophleps magnata, Rhizopogon vinicolor, Montagnea arenaria*
2g. *Bondarzewia occidentalis, Stereum gausapatum, Climacocystis borealis, Xylodon sambuci*

3a. Parasitizing a mushroom or insect, or deforming plant growth (page 90) *64a*

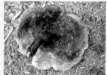

↓3b. Fruitbody fleshy, cup-like to club-shaped (ascomycetes) (page 73) *40a*

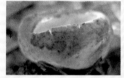

↓3c. Fruitbody brain-like, saddle-like, wrinkled, or ridged (ascomycetes) (page 64) *26a*

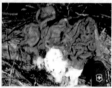

↓3d. With soft, removable sponge under the cap (boletes) (page 48) *5a*

3e. With a stipe; underside of cap smooth, veined, or with blunt ridges *4a*

3a. *Cordyceps militaris, Gymnosporangium* sp., *Hypomyces lactifluorum, Hypomyces rosellus*
3b. *Caloscypha fulgens* (albino), *Cheilymenia fimicola, Ciboria rufofusca, Mitrula elegans*
3c. *Morchella tridentina, Gyromitra montana, Helvella crispa, Helvella dryophila*
3d. *Xerocomellus rainisiae, Boletus regineus, Leccinum discolor, Suillus pungens*
3e. *Turbinellus kauffmanii, Cantharellus formosus, Craterellus calicornucopioides, Gomphus clavatus*

4a. (3e) Thin-fleshed, hollow, underside smooth to wrinkled

Craterellus neotubaeformis n.p.   *Craterellus calicornucopioides*

C. tubaeformis is known from EU; the Cascadia species, *C. neotubaeformis* n.p., is distinct. Distinctive edible, 2–3 in. wide, found under conifers late in the season at all elevations. *C. calicornucopioides*, 1–4 in. wide, is distinctive and present on the east coast of NA and from BC to Central CA in western NA, found in clusters near oaks and sometimes pines, winter to spring. One of the choicest of all mushrooms. Odor is fruity when fresh and cheesy-pungent when dried.

↓4b. Deeply depressed center, vase-shaped, underside bluntly ridged

Turbinellus floccosus (above and at right)

Distinctive, large (2–6 in. wide), and abundant (late summer and fall) in many habitats in NA including Mexico. It is fleshy like a chanterelle and can be rosy to orange in color. *T. kauffmanii* (see lead 3e) is tan. *Turbinellus* species can cause significant stomach problems if eaten.

↓4c . Grows singly; not hollow; flesh stringy like a chicken breast when pulled apart; cap dull egg-yellow, bright egg-yellow, rosy tan, or white in color, bruising orange; ± faint apricot odor

Cantharellus formosus

Cantharellus cascadensis

Cantharellus roseocanus

Cantharellus subalbidus

These four distinctive edible species appear from July through November and are distinguished by cap versus underside color. *C. formosus* (1–5 in. wide) is found at all elevations in Cascadia, usually under conifers. It has a dull egg-yellow cap and underside (see key entry 3e). *C. cibarius* (EU) is a look-alike. *C roseocanus* (1–5 in. wide) is associated with spruce and sometimes hemlock in NA. The cap has a rosy blush when young and the underside is very bright egg-yellow in color. *C. cascadensis* (2–6+ in. wide) has a bright egg-yellow cap and the underside is whitish. Found < 2,000 ft. under both firs and oaks in southern WA and in the OR Cascade Mountains. *C. subalbidus* (2–6+ in. wide), with a white cap and white underside, found at all elevations under conifers and oaks in Cascadia. All are most flavorful before heavy rains.

↓4d. Growing in fused clusters; not hollow; underside tan to violet

Gomphus 'clavatus'

Unique with a lavender underside and tan cap, ≤ 6 in. wide. Very good edible but often riddled with larvae of fungus gnats. Fruits under conifers, late summer and early fall. Related to coral fungi (*Ramaria* spp.). Known from all north temperate parts of the world. The Cascadia material could be a distinct sister species.

4e. In fused clusters; fleshy, blue to black underside *Polyozellus* species

Polyozellus atrolazulinus

Polyozellus marymargaretae

Three Cascadia *Polyozellus* species are commonly called blue chanterelles. *P. purpureoniger* (not illustrated) is so far known from coastal WA, AK, and eastern Asia. The three are challenging to tell apart, requiring microscopy and even DNA evidence to be sure. Found in late summer through fall under spruce and other conifers at elevations above 3,000 ft., in clusters ≤ 12 in. wide. All three are edible but bland. Valued by fabric artists for preparing dyes for silk and wool. Contain compounds of interest to researchers seeking cancer cures, notably stomach cancer. Until 2018, all three were known as *P. multiplex*, a species now known to be restricted to northeastern NA. They are in the order Thelephorales and are not closely related to chanterelles.

5a. (3d) Bolete; viscid cap and/or with a partial veil when young that may leave a ring on the stipe when mature (page 58) *14a*

5b. Non-viscid (or slightly slimy) bolete; lacks both a partial veil and a ring on the stipe *6a*

6a. (5b) Stipe lacking scabers (page 49) *7a*

6b. Stipe with scabers *Leccinum species*

Leccinum insigne

Leccinum ponderosum

Leccinum holopus

Leccinum scabrum

Leccinum species are common in north temperate regions and need critical study, so some names may change. All are considered edible, but many people suffer flu-like symptoms after consuming them. If you do choose to eat them, be sure to cook them thoroughly. *L. insigne* (flesh darkening without first turning pinkish) and *L. discolor* (see key entry 3d [page 44], flesh staining pinkish then smoky) found under aspens and sometimes birch in spring to fall and are very similar. *L. ponderosum* found late summer to early fall above 3,000 ft. under conifers. Caps can reach > 12 in. diameter. Cap flesh does not stain when cut. *L. manzanitae* (2–12 in. wide) is very similar and found associated with manzanita. It bruises pinkish red and then smoky when cut. *L. holopus* and *L. scabrum* are associated with birch in summer and fall. Both have weak staining reactions.

7a. (6a) Boletes with coppery, brown, orange, or red pores when young (page 56) *11a*

↓7b. Stipe ± netted; ± blue bruising; flesh colors various (page 52) *9a*

↓7c. Rapidly turns intensely blue when bruised; flesh yellow or white (page 50) *8a*

7d. Stipe rudimentary to absent; sponge contorted

Boletus subalpinus

Mild tasting, truffle-like species (2–5 in. wide), closely related to *B. edulis*. Develops underground and sometimes found under mountain conifers from about 3,000 ft. to the tree line in Cascadia. Formerly considered *Gastroboletus* species. Edible.

Gastroboletus 'turbinatus'

Eastern NA truffle-like species (1–2½ in. wide) typically develops partly underground under mountain conifers but can be found under low-elevation conifers. Cascadia fungus (*G. turbinatus* var. *flammeus*) may be sufficiently distinct to be renamed, possibly as *Neoboletus flammeus*. A second variety/species may also be present. Indistinct odor and taste. Edibility?
G. ruber was known as *Truncocolumella rubra* and may be transferred to genus *Neoboletus*. It has a thin reddish cap and yellow pores, ± spherical to top-shaped. *G. ruber* is found in the Cascades above 4,000 ft. under mountain hemlock.

8a. (7c) Mild; stipe netted; yellow flesh; intense bluing; under conifers

Butyriboletus abieticola

Found late summer under conifers above 3,000 ft. in western NA. Cap 3–5 in. wide. *B. primiregius*, with a redder, smoother cap, fruits in the spring in the mountains of Cascadia. Both have good flavor. The bitter *Caloboletus frustosus* and *C. rubripes* are often found nearby. All bruise intensely blue.

↓8b. Mild; stipe finely netted; yellow flesh; intense bluing; under oaks

Butyriboletus querciregius (above and at right)

Fruits in the fall and sometimes in the spring below 2,000 ft. in Cascadia. Caps 2–10 in. wide. Firm texture and good flavor. Edible, brown-capped butter bolete, *B. persolidus*, may also be present in the same low-elevation oak habitats. *Caloboletus marshii* (see lead 8d), lacks netting; fruits under oaks during the hot, dry weather of summer; and can taste mild when sampled raw. All four species rapidly blue when bruised.

↓8c. Bitter; stipe ± finely netted; intense bluing; under conifers

Caloboletus frustosus
(= *C. conifericola*)

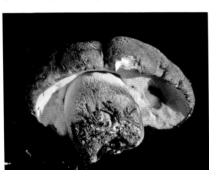

Caloboletus cf. *'calopus'*

Once considered separate species, *C. frustosus* and *C. conifericola* are genetically identical, but the older name, *C. frustosus*, takes precedence. *C. frustosus* usually lacks red on the stipe, and the upper stipe is very finely netted. Both *C. 'calopus'* and *C. rubripes* (not shown) have red on the stipe. *C. 'calopus'* has very distinctive netting, while *C. rubripes* is not netted. (Note: The mushroom I am calling *C. 'calopus'* is not a good match for either of what others have called *Boletus calopus* var. *calopus* or for *B. calopus* var. *frustosus* and might well be an unnamed species.) All three taste bitter when raw, intensely bitter when cooked. Large, stocky, intensely blue-staining, widespread NA boletes can be common under conifers, from August until snow.

8d. Bitter; upper stipe plain; bluing; under oaks; cap 2–6 in. wide

Caloboletus marshii

Fruits in August in hot, dry weather. Can taste mild when raw, but always intensely bitter no matter how it is cooked. Distinguished from the butter boletes by the absence of fine netting on the stipe. Short, stocky stipe, blues intensely. Inedible. Range Cascadia.

9a. (7b) Mushroom bruises blue, but bruising not intense and rapid (page 54) *10a*

↓9b. Rarely slightly blue staining; stipe with a fine to coarse netted pattern; flesh white; tubes white when young *Boletus edulis group*

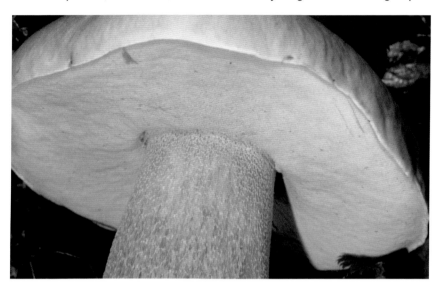

Boletus edulis var. *edulis*

Boletus edulis var. *grandedulis*

Boletus barrowsii

B. edulis var. *edulis* found under spruce > 3,000 ft. from August to October and under spruce on the coast, starting in August in AK and into winter in CA. Coastal *B. edulis* var. *grandedulis* has beige to reddish brown cap, found under shore pines and hardwoods. A white to grayish capped look-alike, *B. barrowsii*, is found inland under conifers and hardwoods. All three are choice and distinctive. All have sponge that turns olive-green in age. All can grow in stupendous abundance. Range NA, EU, Asia, North Africa, Australia.

↓9b. *Boletus edulis group* (continued)

Boletus fibrillosus

Boletus rex-veris

Boletus regineus (above and at right)

Comments: *B. fibrillosus, B. rex-veris,* and *B. regineus* are distinctive species closely related to *B. edulis.* The uncommon *B. fibrillosus* is found under mixed conifers in the fall at 2,500–3,500 ft. Cap surface turns red to pink with application of KOH solution, unlike other *B. edulis* group members. Tubes of *B. fibrillosus* are yellowish when young. *B. rex-veris,* the spring king, is found under mixed conifers in open woods often near grand firs from early April (around 1,000 ft.) through early July (3,000–3,500 ft.). *B. regineus,* the queen bolete, is found in October below 2,000 ft. under mixed conifers where oaks and often pines are present. (*B. aereus* is a EU look-alike.) The caps of *B. regineus* are dark brown but covered with a distinctive powdery white to gray bloom when young (see 3d #2 [page 44]). A gorgeous red-brown–capped *B. rubriceps* (resembles the image of *B. edulis* var. *grandedulis*) is found in the Rockies and can reach > 12 in. diameter. All are choice edibles in Cascadia.

↓9c. Associated with hardwoods; ± viscid; bright yellow pores

Aureoboletus citrinoporus

Cap is cinnamon color to reddish brown, 2–6 in. wide; lower stipe is pallid, colored above. *A. flaviporus* has a cinnamon-colored cap and a brown lower stipe, paler above. While edible, both have soft flesh and an acid taste. Range Cascadia.

↓9d. Velvety cap (2–8 in. wide); in the fall on rotting conifer (hemlock) logs

'Aureoboletus' mirabilis

Fruits in August in the mountains and through November at lower elevations in Cascadia. Cap has a lovely velvety feel. Unique and distinctive, this lemony-flavored bolete is a good beginner's mushroom and a very good edible long known as *Boletus mirabilis*. It is so different from other members of the genus that it may be split off into a genus of its own.

9e. Yellowish stipe with tiny scales, reddish near base; with hardwoods

Hemileccinum subglabripes

Cap to 4 in. wide, chestnut to reddish brown. Pale yellow flesh rarely turns slightly blue when cut. Indistinct odor. Edible. Tastes mild to slightly acidic. Uncommon, usually with birch, sometimes with conifers in NA.

10a. (9a) Cap (2–6 in. wide) olive-buff to olive-yellow, red in age

'Boletus' smithii

Cap is velvety to fibrillose. Stipe always has red present. Flesh may or may not bruise blue. Edible with a mild lemony flavor, unlike the somewhat similar bitter bolete, *Caloboletus rubripes*. *B. smithii* is found at all elevations in Cascadia. This species will be transferred to a new small genus, *Pulchroboletus*.

↓10b. Cap (1–3 in. wide) dark vinaceous black, ± faint bloom

Xerocomellus zelleri   *Xerocomellus atropurpureus*

Both *X. zelleri* (also EU) and *X. atropurpureus* are considered good edibles. The main visual difference between the two species is that the cap of *X. zelleri* is finely velvety when young and always has a lighter whitish beige to yellowish tan band on the cap margin, while *X. atropurpureus* lacks both features. Both have long been known as *Boletus zelleri*. Both found late summer to fall under conifers and hardwoods in Cascadia.

↓10c. Cap (2–3 in. wide) brown and cracked, ± pink in the cracks; on the ground

Xerocomellus diffractus group

Common and widespread, mediocre edible. (*Boletus chrysenteron* is an EU look-alike.) Cascadia has three other species in the group: *X. amylosporus*, *X. mendocinensis*, and *X. rainisiae* (see lead 3d #1 [page 44]). All easily confused with *X. zelleri*. Found under conifers and hardwoods in fall.

↓10d. Velvety cap; pores pale brown when young, pores and flesh ± bluing

Porphyrellus porphyrosporus

Dark brown cap (2–6 in. wide) and stipe (1½–5 in. long). White flesh stains blue, then reddish brown. Odor ± antiseptic. Mild to bitter taste. Edibility? In mixed and coniferous woods, sometimes on wood, in summer and early fall in NA, EU.

10e. Slightly viscid, tomentose yellow cap; on wood chips or near stumps

Buchwaldoboletus sphaerocephalus

Cap 2–4 in. broad, nearly flat in age, often cracked like mud. Pores stain blue, then brownish. Stipe 2–4 in. long, yellow, lacking reticulations. Indistinct odor and taste. Edibility? Image taken in the fall on a sawdust pile. Range NA, EU.

11a. (7a) Fruits under oaks and sometimes other hardwoods *13a*

11b. Fruits most often in coniferous forests *12a*

12a. (11b) Slender; pores coppery to pinkish; stipe base yellow; cap 1–3 in. wide

Chalciporus piperatus group

Common small, peppery bolete is believed to be toxic. Rarely stains blue and the flesh is yellow. *C. piperatus* is suspected of being a mycoparasite on *Amanita muscaria* mycelium. Found at all elevations in NA, EU, and Asia, in the late summer and fall. At least two species in Cascadia.

↓12b. Cap (3–8 in. wide) dark reddish brown; blue staining; red pores

Rubroboletus pulcherrimus

Causes severe gastric distress when eaten, one death recorded. Do not even taste *R. pulcherrimus* or *R. eastwoodiae*. The taste is very unpleasant. Fortunately, it is rare, appearing in late summer and fall in western NA.

12c. Non-bulbous stipe, granular covering; cap blackish to reddish

Neoboletus 'luridiformis'

One undescribed species in Cascadia. The cap surface turns red in KOH, quickly darkening; flesh turns yellow to pale orange in KOH. Cap 2–5 in. wide, stipe 2–4 in. long. Indistinct to unpleasant odor and taste. *Neoboletus* species are under conifers and sometimes hardwoods. Edible. Range NA.

13a. (11a) Medium stature (caps 2–5 in. wide); non-reticulate stipe with reddish granules; red to orange tube mouths; blue bruising of all parts

Suillellus amygdalinus

Common in warmer and drier microclimates of WA and OR, fruiting late September through November under hardwoods. In warmer parts of coastal CA, fruiting can continue into spring. Indistinct odor, mild taste. Edible but not recommended.

13b. Stipe reticulate, bulbous; red tube mouths; blue staining

Rubroboletus eastwoodiae

This infrequent, distinctive, extremely poisonous species was long known as *Boletus satanas*, an EU look-alike. Found below 2,000 ft., October through December under oaks in Cascadia. Large (3–12 in. wide). Odor mild, resembling dung with age. Taste indistinct. Do not even taste it. The taste is very unpleasant.

14a. (5a) Stipe lacking a persistent ring or ring zone (page 61) *21a*

14b. Stipe with a persistent ring or ring zone *15a*

15a. (14b) Flesh of stipe base not staining bluish green
(see also key lead 16a) *17a*

15b. Stipe base staining bluish green within 15–30 minutes *16a*

16a. (15b) Cap (2–6 in. wide) fibrillose-scaly when young, cinnamon

Suillus lakei (above and at right)

Blue staining is weak and only at the base of the stipe. Pores bruise brown. Cap slightly slimy, slightly to strongly fibrillose. Abundant under Douglas fir at all elevations. Appearing in the mountains in late summer in western NA. *S. lakei* var. *pseudopictus* is distinguished by larger scales and a redder cap (no longer considered a separate variety). Edible.

16b. Cap (2–6 in. wide), reddish brown to orange-brown, scattered fibrils

Suillus caerulescens   *Suillus ponderosus*

These species are difficult to tell apart, especially as they age. When young, the veil of *S. ponderosus* has a yellow, viscid lower layer that is absent in *S. caerulescens*. The cap and stipe of *S. ponderosus* may turn green. Both exhibit weak blue staining at the stipe base and elsewhere bruise brown. Both edible. Summer–fall in Western NA.

17a. (15a) Cap viscid (slimy), not woolly-fibrillose-scaly; under various trees *19a*

17b. Cap dry, woolly-fibrillose; associated with western larch *18a*

18a. (17b) Cap (1–4 in. wide) brown scales on yellow; stipe base hollow

Suillus ampliporus

Resembles S. *lakei* but easily distinguished by the hollow stipe (and angular pores). Always near larch in the mountains from late summer until snowfall in Northern Hemisphere. Good edible (for a *Suillus* species). Very distinctive. Long known as S. *cavipes.*

18b. Cap (3–8 in. wide) fibrillose-scaly, rosy red; stipe base not hollow

Suillus ochraceoroseus

Unique and distinctive species was long known as *Fuscoboletinus ochraceoroseus.* It fruits in late summer through fall in western NA mountains, always in association with western larch. It is bitter and thus not suitable as an edible.

19a. (17a) Cap (1½–6 in. wide) viscid; veil viscid; associated with larch

Suillus clintonianus

Glutinous cap and outer veil. The margin of the cap, the partial veil, and the ring it leaves are yellow. Common wherever larch is present in NA, and was long known here as S. *grevillei.* Fruits in late summer into fall. Distinctive. Edible but not desirable.

19b. Associated with other conifers, especially pines *20*

20. (19b) Cap (< 5 in. wide) viscid; veil viscid

Suillus acidus (= S. subolivaceous)   *Suillus flavidus (= S. umbonatus)*

Suillus americanus (= S. sibiricus)   *Suillus luteus*

S. acidus is typically associated with western white pine but can be found under other conifers. It has conspicuous glandular dots on the stipe. It is 2–4 in. wide, fruits late summer to fall, and has a mild odor and slightly acid taste. *S. flavidus* is slimmer and smaller, ≤ 2 in. wide, with indistinct glandular dots on stipe. Fruits under lodgepole and shore pines. Odor indistinct, taste sour. *S. americanus* is associated with western white pine. All parts bruise dull cinnamon to vinaceous brown. Dark brown glandular dots on stipe. Odor indistinct, taste indistinct to acidic or bitter. Some reports of contact dermatitis. *S. luteus* is associated with several conifer species. It is 2–5 in. wide and mainly fruits in the fall. Does not stain. Stipe is glandular dotted, at least above the ring. Mild odor and taste. None of these four species are desirable as edibles. If you choose to eat them, remove the slime layer to minimize episodes of diarrhea. *S. flavidus* and *S. luteus* are widespread.

21a. (14a) Cap (< 6 in. wide) margin with a cottony roll when young

Suillus glandulosipes

Suillus brunnescens

S. glandulosipes has a buff-colored cap when young, aging to orange-cinnamon with a white partial veil that does not leave a ring. Cap of *S. brunnescens* (= *S. borealis*) can be whitish to orange-cinnamon and ages to vinaceous brown with a white to red-brown partial veil. *S. pseudobrevipes* is pale to dark honey-yellow with a whitish partial veil. All mainly found under pines in late summer and fall in temperate NA, and all have glandular dots on the stipes. GI distress possible if eaten. All three are bland edibles.

Suillus pseudobrevipes

21b. Cap margin naked *22a*

22a. (21b) Stipe with brownish glandular dots (dark brownish spots) (page 62) *23a*

22b. Stipe lacking glandular dots; cap (< 4 in. wide) dark red-brown to cinnamon

Suillus brevipes

This non-bruising species is very common under pines in NA. Cap can be much darker reddish brown than shown in the photo. *S. quiescens* may be the actual species shown in photo. *S. pallidiceps*, also under pines, looks like a pale variant, with a white, yellow, or cinnamon-buff cap. These are bland edibles, with no similar toxic species.

23a. **(22a) Flesh bluing when cut; cap (< 5 in. wide) ± woolly**

Suillus tomentosus var. *tomentosus*

Suillus discolor

Both S. *tomentosus* var. *tomentosus* and S. *discolor* are common in the fall under pines. S. *discolor* was long considered a variety of S. *tomentosus* but proved to be sufficiently distinct to be raised to species rank. Surface fibrils of S. *discolor* easily wash off in the rain, and then it resembles S. *tomentosus* var. *tomentosus*. Sometimes these species literally carpet the ground under pines in NA. Edible but sour.

23b. Not bluing when bruised *24a*

24a. (23b) Cap ± bald *25a*

24b. (Cap< 6 in.) very fibrillose-scaly

Suillus cf. *fuscotomentosus*

When older, fibrils are thinner and S. *fuscotomentosus* resembles a non-bluing S. *tomentosus*. Edible but poor flavor. Common in the fall under pines in Cascadia.

25a. (24a) Pores small, ≤ 1 mm; cap (< 5 in. wide) pale yellow to brown

Suillus 'granulatus'

The photo is of eastern S. *'granulatus'*, which is distinct from the EU species. The new NA name is unresolved, but S. *weaverae* is likely. S. *subalpinus* is one proposed name for a species that looks much like this one. They fruit in the fall under pines in Cascadia and are mild-tasting edibles. Sometimes these species are called "milk mushrooms" because milky droplets often appear on the pores.

25b. Pores large, > 1 mm; cap (< 6 in. wide) orange-cinnamon to purplish brown

Suillus punctatipes

Pores are radially arranged and elongated. Fruits late summer through fall under mixed conifers, often white pine in NA. Edible (but bland). On all viscid *Suillus* species, the slime layer should be removed (or a brief spell under the broiler can dry up the slime layer).

26a. (3c) Parasitizing a mushroom, insect, or plant (page 90) *64a*

↓26b. Fruitbody lacking a distinct stipe (page 73) *40a*

26c. Fruitbody with a distinct stipe *27a*

27a. (27a) Head not skirt-like, attachment to stipe at top and lower on stipe *29a*

27b. Head skirt-like, attachment to stipe only at the top *28a*

28a. (27b) Head very wrinkled, < 2 in. wide; stipe > ½ in. wide

Verpa bohemica

Fruits under cottonwood trees when their leaves reach the size of mouse ears (late February into June depending on latitude and elevation). Cap color sometimes yellow to apricot. Poisonous raw, edible cooked. Even when cooked, can cause loss of muscular coordination or GI distress. Found in most north temperate regions.

28b. Head slightly wrinkled, < 1½ in. wide; stipe diameter < ½ in.

Verpa 'conica' (above and at right)

In riparian areas near a wide range of trees and shrubs in spring. Uncommon. Poisonous raw. Edible when cooked. The two unnamed species with and without brown bands on the stipe are distinct from *V. conica* (EU). Look-alikes are found in north temperate regions.

29a. (27a) Fruitbody not resembling a hollow conifer cone (page 70) *36a*

29b. Fruitbody resembling a hollow conifer cone (the morels) *30a*

30a. (29b) Pitted head with ± vertically arranged ridges (page 66) *31a*

30b. Pitted head (1–5 in. wide), ± randomly arranged ridges

Morchella americana (above), *Morchella prava* (below)

M. americana, M. prava, and a *M. prava* look-alike are distinguishable only by DNA. All appear around May, when lilac trees are in bloom. All three sometimes have some reddish staining. In Cascadia, *M. americana* is normally found near cottonwood trees along riverbanks. *M. prava* can be found in orchards, vineyards, and conifer woods. The *M. prava* look-alike is an EU species, *M. vulgaris* (= *M. dunensis*), and so far is confirmed only from White Salmon, WA, though also believed to grow near the Great Lakes. Several *M. americana* look-alikes are in eastern NA. All are poisonous raw. Choice edibles when thoroughly cooked. The *esculenta* clade.

31a. (30a) Morels associated with recent forest fires (page 69) *35a*

↓31b. Disturbance morels (logging, new construction, burn piles, or bark mulch); close ridges and "ladder-like" look

Morchella importuna

Frequently in clusters. Caps < 3½ in. wide. Fruits when trilliums are starting to bloom. Very good thoroughly cooked. Fruits for 2–3 weeks. Has spread worldwide. Many other morel species fruit in bark mulch but lack this distinctive shape.

31c. The natural morels, the *elata* clade *32a*

32a. (31c) Grows in cespitose clusters or singly (page 68) *34a*

32b. Grows singly, never white to cream-colored *33a*

33a. (32b) Stipe nearly reaches top of the cap; cap margin attached with a deep sinus; usually under cottonwoods in May

Morchella populiphila

Morchella populiphila PNW01

Cap of *M. populiphila* (DNA confirmed) is < 1½ in. wide. All parts are fragile like *Verpa* spp. Found along streams. Cap of *M. populiphila* PNW01 (unnamed) is 1–2 in. wide and firm like most other morels. Found in grass and under hardwoods. Both are delicious (different taste). Compare both to *Verpa bohemica*. Range western NA.

↓33b. Stipe ± long; cap attached to stipe with a shallow sinus; cap < 3 in. wide; stipe ± fragile, but not as fragile as *Verpa* spp.

Morchella brunnea

Grows singly under cottonwoods, in mixed forests, and in pure conifer stands when trillium blossoms are aging pink and calypso orchids are blooming in Cascadia. Can be gregarious but rarely cespitose. Poisonous raw. Meaty and delicious when well cooked.

33c. Widespread under conifers and hardwoods and sometimes waste areas; stipe not fragile

Morchella norvegiensis

Synonym *M. eohespera*. Cap < 2 in. wide. If *M. snyderi* is growing singly, these two species can easily be confused. Both are choice edibles when well cooked. Found on both NA coasts and in China, Scandinavia, the Alps. In Cascadia, it is found at all elevations, fruiting as early as late March at sea level to July at 3,500 ft. One of the most abundant morels in many areas of the world.

34a. (32a) Often in clusters; ridges blacken in age; pits variably colored

Morchella snyderi

Pit color varies and includes pale tan, greenish to burgundy, and blackish. Found singly and in clusters in the same areas year after year in Cascadia. Two friends once picked 800 pounds in a day in one ± 40-acre plot. Other friends and I then picked 24 gallons of morels there over a two-week period. Most abundant when the trillium blossoms are turning pink and the calypso orchids are blooming. Very delicious, thick, and meaty mushroom; one of my favorites. Can be > 6 in. tall. Some people are highly sensitive to morels and can get very ill after eating them, even if they have been well cooked.

34b. Sometimes clustered; ridges age ± cream-colored

Morchella tridentina

Also in EU. Fruits in the middle of morel season. Pits start out grayish white and are often pale ochre in age. Sometimes some reddish bruising. Can be ≤ 6 in. tall. Found in the same areas year after year. Synonym *M. frustrata*. Choice! (Also see key lead 3c #1 [page 44].)

35a. (31a) Found 1–2 years after a burn; variable color pits; ridges black in age

Morchella eximia

Several burn morel species that fruit when the trillium blossoms are turning pink and the calypso orchids are blooming are distinguished by DNA. They include *M. eximia*, *M. exuberans*, *M. septimelata*, *M. sextelata*, and one unnamed species. Can be 2–6 in. tall and appear in stupendous abundance after a fire (or even two years after if year one is dry). Can be gritty from ash and may have absorbed fire-retardant chemicals. Otherwise edible and very good. Also known from EU and Asia.

35b. Found after a burn from late June until snowfall; < 5 in. tall

Morchella tomentosa

Densely covered with small black hairs (evident when young, dispersed in age). Buff colored to pale yellow when mature. Appears in June until snowfall if moisture levels permit in NA. Meaty and delicious. Can be abundant. May fruit for several years after fires.

36a. (29a) Head saddle-shaped (page 72) *38a*

36b. Head brain-like or irregularly saddle-shaped *37a*

37a. (36b) Stipe width < cap diameter; round to flattened; cap ± dark red-brown

Gyromitra cf. *venenata*   *Gyromitra* cf. *splendida*

Several look-alike, 2–6 in. tall, NA species fruit in great abundance under conifers, sometimes hardwoods, at all elevations shortly before and during morel season. Though they are very tasty, they are very toxic and dangerous, even when cooked. Toxin levels vary greatly with region, soils, and elevation. The toxin gyromitrin decomposes to the carcinogenic monomethylhydrazine in the stomach or upon cooking. *G. venenata* (hardwoods), *G. splendida* (conifers), and *G. anthracobia* or closely related sister species are the *G. esculenta* (EU) look-alikes found in Cascadia so far. They can cause severe GI distress, severe liver damage, and even death.

↓37b. Stipe convoluted, ± wide as cap (≤ 7 in.); snowbanks

Gyromitra montana

Near conifers as soon as the snow melts, at all elevations. *G. gigas* is a very similar EU species. *G. korfii* and *G. americanigigas* are in eastern NA. *G. brunnea* and *G. caroliniana* are two other eastern NA species with similar convoluted stipes. All are distinctive and edible, but difficult to clean. Compare to the possibly deadly *G. esculenta* group (37a.).

↓37c. Stipe fluted, ≤ head diameter; fruits after most morels are done

Gyromitra californica

Appears under conifers in June and July in the mountains of western NA. Typically, 2–4 in. wide. Stipe may have red bruising, cap may be orange-brown to dark brown. Stipe creamy white to orange-brown. Probably poisonous.

↓37d. Stipe ridged and pitted, longer than head diameter; under conifers

Helvella vespertina

H. lacunosa is a EU look-alike. Head is sometimes saddle-shaped; cap margin attached to the stipe in places, color gray to black. Two similar white to cream-colored species are *H. crispa* (see 3c. #3 [page 44]), and *H. lactea. H. maculata* is similar but has a mottled cap and the margin is not attached to the stipe. All are 2–6 in. tall, found under conifers in fall (sometimes spring) at all elevations. Edible if well cooked.

37e. Stipe ridged and pitted; short stature; mainly under hardwoods in the spring

Helvella dryophila

Fruits before the morels, mainly under oaks but also under fruit trees and other hardwoods; 1½–4 in. tall. At times abundant from central CA to northern WA. Edibility? Sometimes unfolds like a sheet of origami paper when cut from top to bottom through the middle.

38a. (36a) Underside of head with soft hairs

Helvella compressa

Distinguished by copious, short, whitish hairs on underside of cap. Spring species found under conifers, sometimes mixed woods from CA to AK. Edibility? Too small (½–2 in. wide and 1–2 in. tall) and thin-fleshed to try eating.

38b. Underside of cap smooth to whitish and downy *39a*

39a. (38b) Head reddish brown to orange-brown

Gyromitra infula group   *Gyromitra ambigua*

G. infula is < 12 in tall and 5 in. wide. Brittle cap. Fruits on the ground or on rotten conifer wood, under conifers and in mixed woods from spring to fall. *G. ambigua* is darker and more wrinkled. There is one yet unnamed sister species in Cascadia. Toxicity unclear but gyromitrin has not been detected. Both found in NA, EU, Asia.

39b. Head < 1 in. wide, thin-fleshed, gray to brown

Helvella albella   *Helvella elastica*

H. albella has a downy underside when young, while *H. elastica* is smooth. Both found in coniferous woods and mixed woods, in fall or spring in NA, EU, Asia. Global north temperate distributions. Edibility?

40a. (3b) Fruitbody not distinctly cup-like (page 84) *54a*

↓40b. Upward-facing cup; stipe buried, very short, or absent (page 75) *44a*

40c. Upward-facing cup; stipe obvious *41a*

41a. (40c) Stipe highly ribbed

Helvella solitaria

Also known as *H. queletii*. Found in coniferous and sometimes mixed woods in spring through fall in NA, EU. Edibility? Too small (≤ 2 in. wide) and infrequent to be of interest as an edible. Compare to *Dissingia crassitunicata* (Cascadia) and *D. leucomelaena* (a widespread temperate species).

41b. Stipe cylindric *42a*

42a. (41b) Cup ≥ ¾ in. wide

Helvella fibrosa

Also known as *H. chinensis*. Underside of cup and stipe are both finely hairy. Cap is typically ¾–1 in. wide. Occurs spring and fall under conifers and hardwoods and in mixed woods. Small, thin-fleshed. Edibility? Found in all suitable north temperate regions, though there may be a sister species in Cascadia.

42b. Cup < ¾ in. wide *43a*

43a. (42b) Cup ± chestnut-brown

Ciboria rufofusca   Sclerotinia veratri

C. rufofusca is on fir cones while *S. veratri* grows from a black crust or a rice grain–like sclerotium on *Veratrum* (false hellebore) species in Cascadia. (See also key lead 58c [page 86] for additional, similar species.)

↓43b. Inside of cup tan, yellow, or orange; cap margin white-fringed

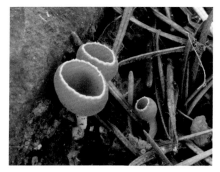 

Geopyxis carbonaria   *Geopyxis deceptiva/G. rehmii*

G. *carbonaria* fruits in the spring in burn areas the year following a fire and always has a distinct stipe. G. *rehmii* (often called G. *vulcanalis*) and G. *deceptiva* are indistinguishable without a microscope. They are more fragile, with or without a stipe, and may or may not be found in burns. All globally distributed.

43c. Fruiting body entirely black; ± long stipe; on wood

Donadinia nigrella (= Plectania nannfeldtii)

Found under conifers and on wood above 3,000 ft. near melting snowbanks in Cascadia. Easily overlooked because of dark coloration and small size (< 1 in. wide). Too small to consider for the table. Edibility?

44a. (40b) Stipe absent or very short, ± round (page 78) *47a*

44b. Stipe ribbed *45a*

45a. (44b) Ribs on a short, largely buried stipe extend well under < 3 in. wide cup

Helvella acetabulum group

Usually found under hardwoods from spring into early summer in NA, EU, Asia. Edibility? *H. costifera* is gray-brown with a densely fuzzy underside and fruits in summer under conifers and hardwoods in NA, EU, Asia. Edibility?

45b. Ribs barely extend under the cup *46a*

46a. (45b) Cup interior dark gray to black; stipe and half of cup below ground

Dissingia crassitunicata

Exterior dark brown to gray-brown above; downy, white below ground. Stipe and lowest parts of cup ribbed. Under conifers in spring (OR to AK). Small, thin (< 2½ in. wide). Edibility? Formerly in genus *Helvella*. *D. leucomelaena* is a widely distributed look-alike not found in Cascadia.

↓46b. Cup or disk (< 4 in. wide), off-white to orange-brown

Gyromitra melaleucoides

Gyromitra ancilis group (> four species)

Gyromitra leucoxantha group
(two species)

Gyromitra olympiana

Under conifers during morel season. All of these species have overlapping features and can be reliably separated only by distinctive mature spore shapes (*G. melaleucoides* non-apiculate; *G. ancilis* group apiculate; *G leucoxantha* group concave apiculi; *G. olympiana* thickened blunt apiculi). Odor and taste of *Gyromitra* species indistinct. All must be well cooked before eating but are best avoided. All except *Gyromitra melaleucoides* are in the *Discina* subgenus. (See also *G. mcknightii* **and** *G. leucoxantha* PNW01 [page 314].) Note: *G. leucoxantha* PNW01 will be named *Gyromitra persicula*.

46c. Fruitbody with a distinct chlorine odor

Disciotis venosa group (several species)

The inside of the cup is distinctly ribbed; the margins are often turned down. This is a close relative of the morels and is edible when well cooked. Distribution EU, NA. (See also *Disciotis* PNW01 [page 314].)

47a. (44a) Cup blue, red, orange, or yellow; on wood (page 82) *52a*

↓47b. Cup white, red, orange, or yellow; on ground (page 83) *53a*

47c. Color of inside of cup lavender, cream, brown, or black *48a*

48a. (47c) Associated with forest fires or fire pits *50a*

48b. Not associated with forest fires or fire pits *49a*

49a. (48b) Cup goblet-like, tulip-shaped, or ± black and shallow (page 80) *51a*

↓49b. Medium to large (< 4 in. wide), fragile cup; on soil in woods

Peziza species

Peziza brunneoatra

Legaliana badia

Peziza varia

Peziza domiciliana

Peziza species are difficult to identify even with a microscope. *P. brunneoatra* grows widespread on damp soils, usually on the edge of the woods. It is small, ≤ 1 in. wide, with brownish flesh. *L. badia* ≤ 4 in. wide, with an olive-brown inner surface and reddish brown flesh. *P. varia* (= *P. repanda* and *P. cerea*) is highly variable both macroscopically and microscopically and can be ≤ 6 in. wide. *P. domiciliana* grows on alkaline surfaces, often indoors, < 5 in. wide. Flesh is yellowish. No *Peziza*-like species are of interest as edibles.

↓49c. Grows on manure piles, well-manured ground, and compost

Peziza vesiculosa

The margin of the < 4 in. wide cup is distinctively incurved, convoluted, and split. Fruitbodies can appear in any season on richly manured ground, often in dense clusters. Not a recommended edible.

49d. Asymmetrical cup, ± 2 in. wide, resembles rabbit ears *Otidea species*

Otidea onotica

Otidea species

Otidea alutacea group

The named *Otidea* species found in Cascadia as of this writing are *O. leporina, nannfeldtii, onotica, oregonensis, propinquata, pseudoleporina, rainierensis,* and *tuomikoskii.* They fruit mainly in late fall under Douglas fir and oaks. The distinguishing pinkish tones of *O. onotica* fade at maturity. Some *Otidea* species may contain the toxin monomethylhydrazine (unconfirmed). Avoid.

50a. (48a) Small (< 2 in.), shallow cup; violet-colored when young

Geoscypha violacea

The other violet-colored burn cup, *G. tenacella* (= *Peziza praetervisa*?) is distinguished by spiny (versus smooth) spores. *Daleomyces bicolor* (page 314) is more intensely violet colored and is not associated with burns. Edibility? None are worth trying. Range NA, EU.

50b. *Peziza*-like burn cups, ± with tufts of hairs on outside

Plicaria endocarpoides   *Plicaria carbonaria*

Spiky, round spores (microscopic) and dark tufts of hairs on the outside of the cup of *P. carbonaria* distinguish it from *P. endocarpoides*, which has smooth, round spores and few, if any, dark tufts of hairs on the outside of the < 3 in. wide cup. Neither is edible. (Note: Spores of *Peziza* species are elliptical in shape.) Range NA, EU.

51a. (49a) Large (2–8 in. wide); buried hollow ball opens like a tulip

Sarcosphaera 'coronaria'

Very common spring species, found at all elevations in NA and EU during morel season, known as the tulip cup. Begins underground as a hollow ball. Whitish interior turns violet as spores mature. Edible, but hyperaccumulates arsenic. The one available Cascadia DNA sequence does not match EU species.

↓51b. Interior cream-colored to grayish; small (< 2 in. wide), deep, often scalloped goblet

Tarzetta 'catinus'/T. 'cupularis'

T. 'catinus' and *T. 'cupularis'* look alike but differ microscopically and by DNA. Both are often associated with disturbed ground, sometimes burns. Their DNA does not match *T. cupularis* (under conifers EU), or *T. catinus* (under both hardwoods and conifers EU). Edibility?

↓51c. Cup interior pinkish to black; small (< 1 in. wide) goblet; round, whitish stipe

Neournula pouchetii

Fruits under conifers during morel season at all elevations in NA, EU, North Africa. Stipe and white portion of the goblet are below ground. Scalloped top is distinctive. Cascadia species may be *N. nordmanensis.* Edible?

↓51d. Black, ± rubbery, < 2 in. wide cup with a stellate margin

Plectania milleri group

Stellate margin and absence of a stipe makes this dark species distinctive. Found under conifers at all elevations during the main morel season in Cascadia. Edibility? There are two closely related species, one with a less distinctly stellate margin (see page 315).

51e. Dark, ± rubbery, < 2 in. wide, non-stellate cup, ± stipe; on wood or ground

 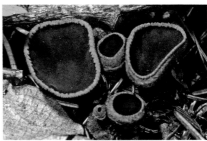

Pseudoplectania nigrella   *Plectania melastoma/P. harnischii*

P. nigrella lacks much stipe. *P. melaena* (= *P. vogesiaca*) is always on wood in coniferous forests and has a short to longish stipe and scanty coiled hairs on the underside. Interior creamy to grayish, with a small (< 2 in. wide), deep, often scalloped goblet. *P. melastoma* and *P. harnischii*, both in Cascadia, have orange granules on the cup margin and exterior and are distinguished microscopically. All four species are found in early to late spring, some globally. Edibility?

52a. (47a) Cup interior red, > 1–2 in. wide; attached to sticks

Sarcoscypha coccinea

Found typically clustered on hardwood sticks (sometimes buried sticks) in winter and early spring. Distinctive. A mediocre edible, rarely found in quantity. Too small to be of interest as an edible. Found west of the Rocky Mountains and in EU.

↓52b. Cup yellow to red, < ½ in. wide

Calycina (= Bisporella) citrina

Scutellinia scutellata group

Lachnellula arida

Capitotricha bicolor

These four species (all shown ± life-size) are commonly found whenever conditions are wet in many parts of the world. *C. citrina* is easily spotted on sticks. Tiny eyelash cups (the *S. scutellata* group) can grow on wood, on other vegetation, or on the ground and are challenging to distinguish. Several *Lachnum* species are found on larch and other conifer sticks. *C. bicolor* (= *Lachnum bicolor*) grows on hardwood branches and berry canes. It is clothed in dense, white hairs.

52c. Whitish to blue species; < ½ in. wide

Chlorociboria aeruginascens

Found on blue-green–stained wood on the ground when conditions are wet. *C. aeruginascens* often has asymmetrical cups, while *C. aeruginosa* (page 315) often has symmetrical cups. The two species are only reliably distinguished by microscopy. Both found worldwide.

53a. (47b) Cup various shades of orange to red (see also key leads 43b [page 74], 52b)

Aleuria aurantia   *Caloscypha fulgens* group

A. aurantia (≤ 4 in. wide) is unmistakable and common in spring and fall, usually on disturbed ground at all elevations (NA, EU). Edible but tasteless. *C. fulgens* (see also key lead 3b [page 44]), ≤ 2 in. wide, fruits in the spring near snowbanks and persists for a few weeks after snowmelt. Cup can be pure white with blue staining or pale orange to reddish orange with green staining (three species known). It is a Douglas fir seed parasite. Causes gastric upset in some people and is tasteless (NA, EU).

53b. Cup (± 1 in. wide) white, stiff brown hairs on the outside

Humaria hemisphaerica

Found July to October on humus and rotten wood. *Perilachnea* (= *Trichophaea*) *hemisphaerioides* is smaller, looks identical, but is associated with burnt ground (and is distinguished microscopically). Neither is of interest as an edible.

54a. (40a) Fruits on the ground (page 87) *59a*

↓54b. Fruits on wood or conifer needles *56a*

54c. Fruits on dung *55a*

55a. (54c) Near late spring snowbanks on pika and other rodent droppings

Byssonectria cartilaginea

A mass of tiny yellow to orange (± 1/16 in. wide) disks grow on a thick, whitish to orangish mat, covering copious rodent pellets. *B. terrestris* is similar in appearance but forms a thin mat where animals have urinated (pellets of dung rarely are present).

55b. On deer and elk pellets and cow pies; numerous tiny species

Cheilymenia stercorea (single elk pellet, ± 10x life-size)

This spectacular fungus was the most striking of five different species on a single elk pellet. The hairs covering the outside are stellate (viewed under a microscope). *C. fimicola* (see 3b #2 [page 44]) has shorter, non-stellate hairs. NA distribution. (See also *Cheilymenia theleboloides* and *Coprotus* cf. *ochraceus* [both page 315].)

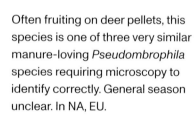

Often fruiting on deer pellets, this species is one of three very similar manure-loving *Pseudombrophila* species requiring microscopy to identify correctly. General season unclear. In NA, EU.

Pseudombrophila cervaria (photo from June, ± life-size)

56a. (54b) Stipe present (page 86) *58a*

56b. Stipe absent *57a*

57a. (56b) Tiny, translucent, yellow to orange disks

Orbilia species

Seventeen documented yellow to orange *Orbilia* species are now described in MycoMatch. The bluish disks are probably *Mollisia cinerea* (= *Tapesia fusca*). These species grow on old wood, often hardwood, and herbaceous stems in NA, EU.

↓57b. Thick disks; on waterlogged wood

Adelphella babingtonii (± life-size)

Found on water-soaked or submerged wood and other vegetation in spring and summer. Color pallid, as shown, to reddish brown. Broadly attached to substrate. Long known as *Pachyella babingtonii*. Global distribution.

57c. Disks (< ½ in. wide); on recently shed true fir needles

Pithya vulgaris

P. vulgaris often appears near melting snowbanks on needles of true firs. The similar but smaller *P. cupressina* (page 315) grows in late winter through spring on recently shed needles of *Juniperus* species and some other conifers. In western NA.

58a. (56a) Brightly colored, convex head (< ¼ in. wide); on waterlogged wood

Vibrissea truncorum group
(± life-size)

Distinctive species can be found on old, waterlogged wood in cold-flowing water in NA and EU, generally only above 3,000 ft. Color ranges from pale yellow to dark red-brown. There may be two species, and one may need a new name. Too tiny to eat.

↓58b. Whitish to ochre-buff colored, convex head (< ½ in. wide); on waterlogged wood

Cudoniella aff. *tenuispora?* and
Pezoloma ciliiferum

The stipe of *C.* aff. *tenuispora* can be pallid or brown to black at the base. The two toothed cups in the image are *P. ciliiferum*. Both species found May to October on waterlogged vegetation. Both are somewhat gelatinous. Too tiny to eat. The look-alike *C. clavus* is found throughout NA and EU (as is *P. ciliiferum*).

58c. Stalked disk (< ¾ in. wide); on downed wood

Tatraea macrospora

Common on hardwoods throughout NA. Formerly in genera *Rutstroemia* and *Calycina* and very similar to other species in those two genera (which are distinguished only microscopically). *Ciboria rufofusca* (see key lead 43a [page 73]) is darker and distinguished by growth on cone scales.

59a. (54a) Fruits on unburned ground (page 88) *61a*

59b. Fruits on burned ground *60a*

60a. (59b) Orange disks (± ⅛ in. wide) with brown on margin; on burned soil

Anthracobia macrocystis

One of several very small, orange disk and cup fungi that carpet the ground the year after a fire. *A. melaloma* and *A. maurilabra* are similarly colored but have distinct downy brown hairs on the underside. Found during morel season. Small. Odor, taste, and edibility not recorded. Found in EU, western NA.

↓60b. Pale yellow-orange disks with exterior hairs

Tricharina gilva (± half life-size)

Fruits April to July on burned ground after mosses have reestablished. The exterior of the < ¼ in. cup is clothed in light reddish brown hairs. *Anthracobia melaloma* is distinguished by a bright orange color, with exterior clumps of brownish downy hair. Odor, taste, and edibility not recorded for either of these two similar species. Distribution EU, western NA.

60c. Yellowish green to brown, thick disks (< ¼ in. wide); black in age

Ascobolus carbonarius

Surface of fruitbody soon dotted with the black tips of protruding asci; eventually surface is nearly all black. Found spring through summer after a fire. The orange disks in image are *Anthracobia macrocystis* (shown ± half life-size). Global distribution.

61a. (59a) Colors pale to brown or black *63a*

61b. Brightly colored *62a*

62a. (61b) Bright yellow to orange, stalked club (1–4 in. tall); in swamps

Mitrula 'elegans' (life-size)

Widely distributed in swamps at all elevations in NA and EU (three distinct species?). Grows on rotting vegetation. Fruits April through September on rotting submerged vegetation. Very small and tasteless. Only one *Mitrula* species is confirmed from Cascadia region. Odor and taste not reported.

↓62b. Bright yellow-orange, clavate to spathulate club

Neolecta vitellina

Found August through October in needle beds and moss carpets. At ≤ 2 in. tall, you might be tempted to eat it, but the edibility is unknown. It is odorless and tasteless. Found in NA, EU.

62c. Pale tan to yellowish, spathulate club (½–4 in. tall)

Spathularia flavida

Common under firs, pines, and hardwoods, mainly below 2,000 ft. in the fall. At ≤ 4 in. tall, it has been tested for edibility and found tasteless and tough, but not gelatinous. Found in NA, EU.

63a. (61a) Cream-colored to light brownish, convex to convoluted (< 1 in. wide) head; pallid stipe

Cudonia circinans

Fruits May to October under conifers and hardwoods. Can be gregarious. Common under coastal Sitka spruce. One unconfirmed study reported they contained high levels of the carcinogenic liver toxin monomethylhydrazine. Odor and taste not recorded. Poisonous, potentially deadly?

↓63b. Pinkish cinnamon, convex to convoluted head; grayish brown stipe

Cudonia monticola
(= Pachycudonia monticola)

Can be ≤ 4 in. tall and 1 in. wide. Has a solid, fibrous stipe. Fruits mainly on spruce needles and conifer debris. Found above 3,000 ft., next to snowbanks in winter through spring and summer. Odor, taste, and edibility are unknown. DNA results indicate it fits in genus Cudonia, and Cudonia species may belong in genus Spathularia.

63c. Black clubs, ± 3 in. tall

Trichoglossum, Geoglossum, and Glutinoglossum

This species, a black-earth tongue, which was once in the genus Geoglossum, is covered in short, stiff hairs. The actual Cascadia species may be distinct. It is a fall species mostly associated with sphagnum bogs but also found on well-rotted wood and humus. Geoglossum and Glutinoglossum species share the same habitat but lack the hairiness. Odor and taste not recorded. Edible but very small and very tough. Little is known about edibility of the others. Found in NA, EU.

Trichoglossum 'hirsutum'

64a. (3a) Parasitizing another fungus (page 92) *65a*

↓64b. Clubs < 3 in. tall; growing from buried butterfly pupae

Cordyceps militaris   *Cordyceps washingtonensis*

These distinctive, rarely noticed fungi grow in late summer and fall on both butterfly and moth pupae. *C. militaris* is currently cultivated as both a medicinal and an edible mushroom and grows in NA and EU.

64c. Fungal parasites on plants (many different fungi)

Ustilago maydis

When young, infected corn kernels swell and produce a delicacy called *huitlacoche* in Mexico (*U. maydis*, or corn smut). Global distribution? If harvesting your own *huitlacoche*, use some care because corn can also be infected with the superficially similar *Claviceps gigantea*, which can cause ergot-like symptoms (see *Claviceps purpurea*).

Claviceps purpurea

Grows on grasses and grains globally. A dark purplish sclerotium is produced in place of a seed. If the sclerotium, called ergot, is consumed, can cause madness and burning pain (Saint Anthony's fire).

64c. (continued). Fungi that cause distorted plant growth

Taphrina occidentalis

Taphrina deformans

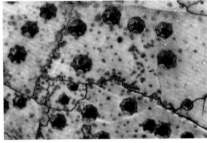

Coccomyces dentatus

Rhytisma arbuti

Cronartium ribicola

T. occidentalis causes a deformed growth of alder cones (alder tongues), which are plant material on which the spores are produced (there is no fruiting body). The young growth is reddish. The tongues overwinter and turn dark brown. T. deformans (peach leaf curl) causes peach tree leaves to redden, with blisters and spores produced on the surface of the blisters that spread by rain. Can weaken the tree enough to kill it. C. dentatus, one of several local *Coccomyces* species, is a beautiful little fungus (visible with a hand lens) infecting Oregon grape leaves. Many different *Rhytisma* species affect leaves of trees and shrubs, but they usually do not cause enough leaf damage to damage the tree appreciably. R. arbuti (tar spot fungus) is found on many different species of trees and shrubs, and R. punctatum is abundant on fallen bigleaf maple leaves. C. ribicola (white pine blister rust) causes extensive loss of white pines. The alternative hosts are *Ribes* spp., especially gooseberries.

65a. **(64a) Parasitizes species of *Russula* and *Lactarius*; fruitbody firm**

Hypomyces lactifluorum

Known as the lobster mushroom, this species mainly attacks the *Russula brevipes* group. Gills and basidiospores never form. Can be white to orange (the edible stages), aging deep reddish, valued as a dye mushroom. Good texture, mild flavor. Generally edible, little inherent taste, good at taking up flavors. Found worldwide.

Hypomyces lateritius

This species attacks a wide range of *Lactarius* species but visibly obliterates only the gills, leaving the host extra firm and identifiable. The affected gill surface is initially white to lemon-yellow, then turns yellowish orange, and finally turns brick-red to red-black. Uncommon, unlike the lobster mushroom that can be abundant. The edibility is unknown, although in the photo, the host was an edible species, *L. rubrilacteus*. It should not be eaten. Circumglobally distributed.

Hypomyces 'luteovirens'

Attacks species of *Russula* and sometimes *Lactarius*, initially turning the gills and stipe (and sometimes the cap) bright yellow, then green, and finally blackish green. It should not be consumed. Found in all north temperate regions. The Cascadia species is distinct from EU species and may be *H. macrosporus* (AL).

↓65b. Attacks boletes, turning them into fetid mush

Hypomyces chrysospermus   *Hypomyces microspermus*

These bolete eaters are distinguished by microscopy. An initial white mold is followed by a yellow to golden stage. The final red-brown, pimpled stage is rarely seen, as the mushroom is gone by then. Edible? No way! Global distribution.

↓65c. Attacks *Helvella* species, turning them into mush

Hypomyces cervinigenus

This mold starts off white and then turns a beautiful pinkish brown and finally cocoa-brown. Frequent on *Helvella vespertina* in late fall and winter; image shown on *H. elastica* in the spring. Not edible. Found in all north temperate regions.

65d. Forms a white, cottony growth, covering many different species

Hypomyces ochraceus (white, asexual stage on *Cortinarius* sp.)

The buff- to orange-pimpled sexual stage is rarely seen. Also rots species of *Cantharellus*, *Russula*, *Lactarius*, and possibly other genera. Probably globally distributed.

66a. (2g) Fruitbody on wood (page 98) *73a*

↓66b. Fruitbody leathery; on the ground (page 98) *72a*

66c. Fruitbody soft, at least when young; on the ground *67a*

67a. (66c) Cap not distinctly scaly *69a*

67b. Cap distinctly scaly *68a*

68a. (67b) Growing from a buried sclerotium; cap scaly ≤ 6 in. wide

Polyporus decurrens/P. mcmurphyi

P. decurrens (Canadian tuckahoe) is mainly found under aspens NA. P. mchurphyi (shown in image) is distinguished only by DNA and is known from hardwoods in CA, WA. Odor of anise or cheese, taste mild and cheesy. Edible when young, but cook thoroughly. NA species are distinct from similar EU species, P. tuberaster.

68b. Not growing from a buried sclerotium; cap ≤ 12 in. wide (see also key lead 83d [page 106])

Scutiger (Albatrellus) pes-caprae   *Scutiger (Albatrellus) ellisli*

S. pes-caprae has a dark brown cap, while S. elisii has a sulfur-yellow to yellow-brown cap with darker brown stains in age. Pores of either may stain green. Found under conifers above 3,000 ft. in late summer and fall. Both are edible, even choice, when young and have a mild odor and taste. Found in NA, EU.

69a. (67a) Multiple (sometimes single) caps from a central to lateral stipe *71a*

69b. Single cap on central to lateral stipe *70a*

70a. (69b) Cap (< 8 in. wide) velvety, brown to gray-brown

A fall species found near old stumps and conifers, probably growing from tree roots. Cap and stipe hairy, but not scaly. Odor ± like iodine. Very bitter and not edible. The color of these specimens is much like that of *Scutiger pescaprae*. Found at all elevations in NA, EU.

Jahnoporus hirtus

70b. Cap smooth to fibrillose, gray to black *Boletopsis grisea* group

Boletopsis (unnamed)

Boletopsis grisea

Boletopsis aff. *leucomelaena*

Unlike other polypores, the tube layer can be peeled off. Unlike boletes, the tube layer is always thin. The most common is *B. grisea* (= *B. subsquamosa*), a white to pale gray, < 8 in. wide species associated with pines. *B.* aff. *leucomelaena* has a black cap and is associated with spruce. The Cascadia species is genetically distinct from the EU look-alike *B. leucomelaena*. All are ± edible, ± bitter tasting.

71a. (69a) Broad cap; rooting stipe

Bondarzewia occidentalis

B. mesenterica and *B. montana* are similar EU species. This large (≤ 8 in. wide) species fruits under larch, true firs, and possibly other conifers. The odor is pleasant, and the taste is mild when young, and tough and bitter when old. Found in EU, western NA.

↓71b. Fused ochraceus, blue, or blue-gray cap (< 10 in. wide)

Albatrellopsis flettii   *Albatrellopsis confluens*

Formerly in genus *Albatrellus*. Widespread in late summer and fall under conifers. *A. confluens* is a EU species that is never blue. The western NA mushroom has a matching ITS DNA region but starts out deep blue and fades to an ochre color. *Neoalbatrellus subcaeruleoporus* (CA to BC) and *N. caeruleoporus* (eastern NA) is smaller, thinner, and often has blue pores. The odor and taste are mealy. It is edible but tough.

↓71c. White, gray-brown, or pale orange to tan cap *Albatrellus* species

Albatrellus ovinus   *Albatrellus avellaneus*

A. ovinus is white to gray-brown (≤ 8 in. wide), stains yellow. Found under conifers and in mixed woods. *A subrubescens* (no photo, caps ≤ 5 in. wide) is similarly colored to *A. ovinus* and bruises yellow (to orange), developing rosy stains in age. Found mainly with two- and three-needle pines. Both are circumboreal. *A. avellaneus* (western NA) is pale orange to tan with dark, fine scales (colors overlap with the other species and a microscope may be needed), caps ≤ 5 in. wide. Found mainly with hemlock and spruce. All three species fruit in late summer through fall. Edible but tough.

↓71d. Large orange to greenish, rosette-like caps (≤ 12 in. wide) age to rusty brown

Phaeolus schweinitzii group (above and at right)

P. schweinitzii is a complex of species, with two unnamed species in western NA. Known as the dyer's polypore, they are valued for their beautiful dye colors on silk and wool. Brown bruising. A major cause of butt rot in Douglas fir. Usually terrestrial but attached to tree roots. May also grow directly on conifers. Also sometimes attacks hardwoods. Sour tasting and suspected poisonous. First image is of young material. Turns black in old age and overwinters, but is no longer useful as a dye mushroom. Distributed circumglobally.

71e. Olive-green to ochre-colored cap (< 8 in. wide); large pores turn purple-gray in age

Polyporoletus sylvestris

Cap fades to brownish. A very rare species related to *Albatrellus* spp. and found from 4,000 ft. to tree line under conifers, probably attached to roots. Unusual spores (microscope). Western NA. Edibility?

72a. (66b) Cap < 4 in. wide, pale cinnamon to deep brown, grayish in age

Coltricia cf. *perennis*

Once known as *Polyporus perennis*, this annual has a velvety, tomentose, strongly zoned cap. It is unclear if the Cascadia species is a genetic match to the EU species. Found in both deciduous and coniferous woods, often on exposed soils along trails in summer and fall. The stipe color matches the dominant cap color. Not edible.

↓72b. Cap (< 2 in. wide), brown to deep reddish brown

Coltricia aff. *cinnamomea*

Previously known as *Polyporus cinnamomeus*, this annual has a silky-shiny, velvety brown to reddish brown cap. The pores are yellow-brown to red-brown and the stipe is dark brown. Usually in deciduous woods in fall. The Cascadia species is an unnamed look-alike to the EU species. Not edible.

73a. (66a) Stipe absent *75a*

73b. Stipe present *74a*

74a. (73b) Radially elongated, ± large pores; cap margin with fringed hairs

Lentinus brumalis (= *Polyporus brumalis*)

Lentinus arcularius (= *Polyporus arcularius,* proposed change)

P. brumalis (mainly on birch, fall to spring, rare in west) is now in the genus *Lentinus*. Some, but not all, authorities have also moved *P. arcularius* (on conifers and hardwoods, summer to fall) to the genus *Lentinus*, where it belongs. Fungoid odor, mild taste, though both are not edible. Both widespread species are circumpolar. Found on aspens and other hardwoods. (See also *Polyporus decurrens* [page 94].)

74b. Pores small, round to radially elongated

Picipes badius

Cerioporus varius group

Picipes tubaeformis

Picipes cf. *melanopus*

All four of these circumpolar species have a dark brown to black lower stipe. P. cf. *melanopus* is distinguished from the other three species by always appearing to grow on the ground but in association with wood. The other three grow directly on wood or woody debris. P. *badius* is usually on logs and stumps of either hardwoods or conifers in summer and fall, rarely on twigs and small branches. Both P. *tubaeformis* (on hardwoods) and C. *varius* (on hardwoods or conifers) grow on twigs, branches, and stumps, but not logs. None of these species has a distinctive odor or taste. All are too tough to eat. P. *tubaeformis* previously has been considered both a variety of P. *badius* and a variety of C. *varius* but is a separate species. There are several distinct *Cerioporus* species, and image #2 shows an unnamed sequenced collection. It is likely that more species remain to be uncovered in the P. *badius* group as well.

75a. (73a) Fruitbody a crust; sometimes forms a thin cap (page 100) *76a*, see also (page 114) *92a*

75b. Not as above (page 102) *77a*

76a. (75a) Stipe absent, underside with tiny white pores; cap < 3 in. wide, ⅛ in. thick and protrudes 1 in., on hardwoods

Trametes versicolor (turkey tail)

Cap is always banded with variable sharply contrasting colors of blue, orange, brown, white, gray, or (from algae) green. May form semicircular caps, as shown, or the pore surface may grow crust-like on wood, with the top part turning outward as a thin, multicolored shelf. On hardwoods. Not edible, but a potential medicinal mushroom. Initial clinical trials have been promising for use as an adjunct (not sole) therapy for treatment of some cancers, including breast and prostate cancers. The taste is supposedly pleasant, but that has not been my experience. It is too leathery to consume directly but can be softened in a pressure cooker or used to make a tea. Circumpolar distribution.

↓76b. Stipe absent; underside smooth, orange; on hardwoods

Stereum hirsutum

False turkey tail with thin, ± 1 in. wide cap has a smooth, orange underside. It can also grow like a crust with a thin top edge turning out in a shelf-like fashion. *S. gausapatum* (see lead 2g #2 [page 43]) is a look-alike, common on oaks, and is distinguished microscopically. See also *S. ochraceoflavum* (page 315). All three are global as are many hard-to-identify *Stereum* species. Nontoxic.

↓76c. No stipe, underside with small, almost tooth-like, violet pores

Trichaptum aff. *abietinum*   *Trichaptum* aff. *biforme*

T. aff. *abietinum* resembles a small, shaggy turkey tail but has ragged, violet pores on the underside. It grows on conifers, rarely on hardwoods. The similar, violet-toothed species, *T.* aff. *biforme*, is found on hardwoods. The DNA of Cascadia species does not match either *T. biforme* (EU) or *T. abietinum* (EU). Odor and taste? Both are inedible.

↓76d. Underside smooth to slightly wrinkled, violet

Chondrostereum purpureum

Flat or with leathery, hairy to hairless caps, projecting ≤ 1 in., in overlapping layers. Upper surface indistinctly zoned and buff-colored with a paler margin. On hardwoods, rarely conifers, year-round. Inedible. Circumglobally distributed.

↓76e. Membranous-leathery, tomentose brown patches on wood

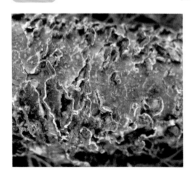

Hydnoporia tabacina

Golden brown, zoned, tomentose caps, if present, resemble small turkey tail fungi. Annual but seen all year since new fruitbodies grow on old ones. On many species of hardwoods, sometimes conifers. Formerly *Hymenochaetopsis tabacina* and *Hymenochaete tabacina*. Inedible. Global distribution.

↓76f. Orange, pink, ± ⅓ in. wide disks; underside of conifer branches

Aleurodiscus grantii
This little cup-like, white-fringed disk resembles an ascomycete but is related to the genus *Stereum* (a basidiomycete). Found on the underside of the lower branches of conifers, especially true firs. Tiny and tough. Western NA.

76g. Tomentose, unzoned, overlapping caps; smoky gray pores

 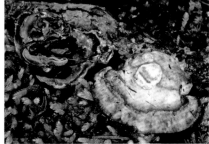

Bjerkandera adusta (young and old)

On hardwoods, especially aspens, rarely conifers. Leathery, often crust-like, with < 2 in. long shelving tops < 1 in. wide extending ≤ 8 in. on logs. An annual, but can be found year-round. Fungoid odor, sour taste. Inedible. Circumglobal distribution.

77a. (75b) Fruitbody hard or gelatinous, even when very young (page 107) *84a*

77b. Soft but not gelatinous when young, may be tough in age *78a*

78a. (77b) Pore surface not covered with a membrane when young *79a*

78b. Pore surface covered with a membrane when young; cap < 4 in. wide

Cryptoporus volvatus
Numerous on trunks of dying conifers. Soft, pleasant-smelling when young, but soon tough. It attracts boring beetles that chew through the lower membrane, eat the pores, and fly off covered with spores that infect new trees. Found in NA, East Asia.

79a. (78a) Fruitbody a shelf-like bracket or shingled, not sponge-like *81a*

79b. Sponge-like fruitbody develops on wood under snowbanks *80a*

80a. (79b) Fruitbody white to tan, bruising darker, shape highly variable

Spongiporus (Postia) leucospongia

Caps ≤ 4 in. wide and 2 in. tall. Feels similar to fine chamois and compresses like a sponge. *Spongiporus* is a segregate genus from the formerly more inclusive genus *Postia*. Found on conifers and aspens in the mountains where snowbanks persist until early summer. Distribution western NA, Himalayas.

80b. Fruitbody orange; sometimes a sheet covering underside of log

Pycnoporellus alboluteus

Develops on logs under snowbanks that persist into early summer. Unique and distinctive, can cover an area ≤ 4 ft. wide. Placing a drop of KOH on it produces a cherry-red color. Usually on conifers but can grow on hardwoods. Sponge-like texture. Inedible. Distribution western NA mountains.

81a. (79a) Fruitbody soft with tan, yellow, or orange coloration (page 105) *83a*

↓81b. Fruitbody soft with red, pink, blue, violet, or black colors (page 104) *82a*

81c. Fruitbody soft, white when young, ≤ 4 in. wide and 1 in. tall *Calcipostia guttulata*

Calcipostia (Postia) guttulata

Short-lived annual. Single to shingled on conifers (especially spruce) and hardwoods. Often has small circular depressions on cap. Bitter and inedible. *Amaropostia stiptica* (= *P. stiptica*) has a rough cap, often with small black dots. *Tyromyces chioneus* (two genetic species in Cascadia) is thin and white but has a cheesy texture and grows on hardwoods. All are inedible and distributed in NA, EU.

82a. (81b) Upper surface a combination of white and blue with short black hairs; < 3 in. wide by 1 in. tall; on old conifer logs

Cyanosporus aff. *caesius*
(= *Postia* aff. *caesia*)

Of twenty-four closely related species in the *Cyanosporus caesius* complex, five thin, ± smooth, ± white species and this one beautiful white and blue species with copius black hairs are in Cascadia. This common thick Cascadia look-alike to *C. caesius* (= *P. caesia*) EU needs DNA confirmation.

↓82b. Upper surface pale pink to pale purple; soft flesh bruises pink

Leptoporus mollis

Fruitbody ± 2½ in. long by 1½ in. thick and 1½ in. tall. Found on many conifers, especially pine and spruce, from summer to fall. Can grow flat on the wood's surface with only the pores exposed, or can be shaped like a thick half-circle. Distinctive. Edibility? Circumpolar distribution.

↓82c. Flesh red and juicy, like raw beef; stipe normally absent

Fistulina hepatica

Can be ≤ 12 in. wide and 2½ in. thick. Very distinctive species is usually half-circular to kidney-shaped. The cap surface varies from gelatinous to velvety, and the color varies from orange-buff to dark red. From southern OR to CA and transglobal, in temperate hardwood forests. Sour taste.

82d. Dark brown to black, velvety; ≤ 6 in. long, 5 in. wide, and 1 in. thick; pores bruise dark

Ischnoderma benzoinum

I. resinosum grows on hardwoods and *I. benzoinium* grows on conifers. Exudes resin. Rotten wood infected with this species smells of anise. Said by some to be edible when young and watery. Circumglobally distributed.

83a. (81a) Orange shelving fruiting body (< 8 in. long, 6 in. wide, 1 in. thick); on hardwoods

Laetiporus gilbertsonii

This coastal (eastern and western NA) species was long known as *L. sulphureus* (eastern and midwest NA). *L. gilbertsonii* has an inferior flavor. When still edible and tender, the margin and pores are yellow. White and chalky in age.

↓83b. Orange fruiting body (≤ 10 in. wide, 6 in. tall, and 1 in. thick); on conifers

Laetiporus conifericola

Common in western NA. Young, tender specimens at the prime age for eating are shown in the image—this is what young *L. gilbertsonii* or *L. sulphureus* look like. White and chalky in old age. The tender margin is harvested for food. Edible. Both species are noted for causing gastric upset in sensitive individuals who eat them, so much so that I no longer eat them.

↓83c. Bright orange-red shelf fungus, ≤ 5 in. wide, 3 in. tall, and 1½ in. thick

Trametes cinnabarina

Spectacular and unique fungus grows worldwide on hardwoods and rarely on conifers. Annual generally found from late summer through most of the winter. No appreciable odor or taste and is not edible. Circumglobally distributed. Long known as *Pycnoporus cinnabarinus*.

↓83d. Annual fruitbody ≤ 7 in. wide; on the ground or shelving on a tree

Onnia tomentosa

Soft, velvety cap yellow-brown to brown. Appears in summer and fall under conifers, often spruce. Often has a stipe attached to the tree roots, but can also be shelving on the tree. Fragrant odor and mild taste, but too tough to eat. Distribution NA, EU.

↓83e. Underside ragged, somewhat maze-like; fruitbody < 3 in. wide, 4 in. tall, and 1 in. thick

Pycnoporellus fulgens

Can be bracket-like (shown) or shelving. Pores are pale orange when young and flesh turns red in KOH. Mainly on conifers. Soft when young but not sponge-like (see *P. aloboluteus*, key lead 80b [page 103]). No odor or taste. Distribution NA, EU.

83f. Cap (< 6 in. wide), tan, bruising dark red-brown, soft annual; above 3,500 ft. on conifers, summer in NA

Amylocystis lapponica

Climacocystis borealis (circumpolar, key lead 2g #3 [page 43]) is distinguished by a soft, sappy, coarsely hairy, white to cream, non-staining fruitbody. Both are mild to bitter. Not edible.

84a. (77a) Fruitbody a gelatinous to hard crust; ± shelving (page 114) *92a*

↓84b. Perennial fruitbody; on wood (page 109) *87a*

84c. Annual fruitbody; on wood *85a*
(see also key leads 90a [page 111] and 91a [page 112])

85a. (84c) Cap surface has varnished look, red to mahogany-red,
sometimes yellow to orange, small white pores

Ganoderma oregonense

Ganoderma tsugae

Ganoderma lucidum group

Ganoderma species

G. *oregonense* (western NA, with two or three pores per mm) and G. *tsugae* (eastern NA, with five or six pores per mm) are found on conifers, usually true firs and hemlock. Fruitbodies ≤ 40 in. wide, 16 in. tall, and 8 in. thick. The G. *lucidum* (reishi) group members are distinguished by growth on hardwoods and occasionally conifers, but they grow near the ground in eastern NA. They can be yellow and orange or mahogany in addition to the red color illustrated. G. *polychromum* (western NA) grows on hardwoods, and the flesh is off-white as opposed to brownish for the G. *lucidum* group. All produce brown spores that often accumulate on the cap (top photo). All can grow with or without a stipe. None is toxic. All are of interest as medicinal mushrooms.

↓85b. Pores ± round, white to buff; fruitbody < 3 in. wide, 1½ in. tall, and ¼ in. thick

Trametes ochracea

Resembles a pale, thick turkey tail. Cap surface whitish to brown with bands of darker brown or reddish brown. Can be greenish with algae. On hardwoods, rarely conifers. Thick, rigid flesh. No odor or taste. Not edible. Distribution NA. DNA data on Cascadia collections not available as of 2023.

↓85c. Pore walls thick and maze-like; fruitbody < 4 in. wide

Cerrena unicolor

Upper surface is coarsely hairy, pale brownish to gray, often greenish from algae. Distinctive black line in the flesh (when cut open). Grows on wounds on living and dead hardwoods, rarely conifers. Inedible. Distribution NA, EU, Asia.

85d. Underside white to brown; pores radially elongated; ± gill-like *86a*

86a. (85d) Whitish pores; white to ochre, bumpy cap with wavy margin

Trametes gibbosa

An annual on hardwoods. Fruitbody < 8 in. wide, 6 in. tall, and 1½ in. thick. Strongly fungoid odor and ± bitter taste. Too tough to eat. Shown growing on cottonwoods along the Columbia River. Common wherever birch grows.

↓86b. Tan pores; cap zoned; fruitbody < 5 in. wide *Trametes betulina* (= *Lenzites betulinus*)

Trametes betulina

Velvety or hairy with concentric zones of tan, gray, brown, or orange, greenish in age from algae. Hard and inedible. Circumglobally distributed, usually on hardwoods.

86c. Ochre to rusty brown pores; radially wrinkled cap, 1–5 in. wide

Gloeophyllum sepiarium

Annual to perennial species on dead conifers and occasionally hardwoods. Margin is usually yellow to orange. Turns the wood red. Little or no odor or taste and too hard to consider for food. Distributed circumglobally, including Africa, with some variability.

87a. (84b) Pores white to brown; ± brown-staining (page 110) *88a*

87b. Pores pink; fruitbody 1–5 in. wide, < ¾ in. thick

Rhodofomes cajanderi (= *Fomitopsis cajanderi*)

On dead conifers and sometimes on hardwoods. Can be shelf-like, bracket-like, or hoof-like. Found year-round in NA, the Caribbean, Central America. *Fomitopsis rosea* (a typically hoof-shaped fungus), now officially *Rhodofomes roseus* (NA, EU, Asia), > 1 in. thick and has paler pores. Neither is edible.

88a. (87a) Pores small and white, fruitbody hard and perennial *89a*

↓88b. Pores maze-like, off-white; fruitbody < 8 in. wide and 3 in. tall

Daedalea quercina

Distinctive and very uncommon species grows on dead and living oaks. Also grows on chinquapin. Tough, corky texture; pleasant fungoid odor; sharp taste. Easy to recognize. Too tough and woody to eat. Circumglobally distributed.

↓88c. Pores brown; cap (< 8 in. wide by 6 in. tall) cracked, brown to blackish bands

Phellinus tremulae

Found on living aspens, *P. tremulae* is a member of the *P. igniarius* group, which includes fifteen species worldwide. Perennial fruitbody can be shelving, hoof-shaped, or bracket-like. Year-round on living hardwoods and dead trees, mostly birch and aspen. Fungoid odor, sour taste, not edible.

88d. Pores tan, aging darker; cap (< 6 in. wide) banded gray, brown margin

Fomes fomentarius group

Grows on hardwoods, especially birch. Pleasant odor, bitter taste, inedible. Known as amadou, the tinder conk (for fire starting). Insides can be felted to make hats, purses, and other items. Currently comprises numerous species. Cascadia species is unclear. Distribution NA, EU, Asia, Africa.

89a. (88a) Found primarily on hardwood (page 112) *91a*

89b. Found on conifers, rarely on hardwoods *90a*

90a. (89b) Upper surface brown to black, rough, ± narrow, and shelving

Heterobasidion annosum group

Always close to the ground. Causes tree death and blowdown. Annual or perennial on living or dead hardwoods but usually on conifers. Can grow flat without a cap. Cap (< 4 in. wide) is rough and irregular, ± covered with fine hairs. Worldwide distribution. Western NA has *H. occidentale* (common) and *H. irregulare* (rare).

↓90b. Upper surface grayish to brownish, with ochraceous band

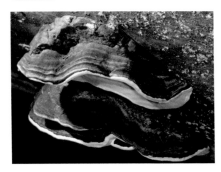

Fomitopsis ochracea

One of two species formerly known as *F. pinicola* (EU) found on hardwoods and conifers. Can be ≤ 9 in. wide, 12 in. tall, and 5 in. thick. A match held to the margin will give a charring reaction. No odor or taste. Woody. Distribution NA. *H. annosum* always grows low on conifers and has a rough top surface.

↓90c. Upper surface with shiny reddish orange marginal band

Fomitopsis mounceae

Long known under the misapplied name *F. pinicola*, a EU species. Can be ≤ 7 in. wide, 6 in. tall, and 3 in. thick. In age, the red belt fades to ochre. On all tree types. Distinguished from *F. ochracea* by holding a match to the margin, which will melt, not char. Distribution NA.

↓90d. Chalky white, crumbly; ≤ 12 in. wide, ≥16 in. tall

Laricifomes officinalis
(= *Fomitopsis officinalis*)

Known as agarikon, of medicinal interest, and was carved as a talisman by coastal Native American shamans. Flesh is cheesy when young, chalky when old. Odor is farinaceous, taste is very bitter. Perennial on living and dead western larch, Douglas fir, and other conifers. Distribution NA.

90e. Rough, shelf-like to bracket-like or flat fruitbody with angular to maze-like pores

Porodaedalea 'pini'

Can be ≤ 8 in. wide and ½–6 in. thick. Very tough and woody. Reddish brown to blackish brown. No odor or taste. Perennial on living older conifers, rarely on hardwoods. *P. pini* is a EU species, and the presence of *P. pini* (versus a sister species) in Cascadia is not confirmed. (See also *P. chrysoloma* [page 316].)

91a. (89a) Margin of cap overhangs pore surface (unique); only on birch

Fomitopsis betulina
(= *Piptoporus betulinus*)

Very distinctive fungus is nearly round in outline, reaching ≤ 10 in. wide and 4 in. thick. Tough and corky when young, hard in age. Fruits spring through fall. Probably annual. Pleasant odor, somewhat bitter taste. Edible but tough.

↓91b. Flattened bracket dusted with brown spores; tubes stain brown

Ganoderma applanatum

Known as the artist's conk because you can draw an image on the white pore layer, which instantly and permanently stains dark brown. Each year adds a new pore layer. Grows mainly on hardwoods but can appear on conifers in the Pacific Northwest. *G. brownii* is similar, less common, and is only on hardwoods. *G. applanatum* can live more than 50 years, 30 in. wide and 16 in. tall.

↓91c. Known only from the base of oak trees and always close to the ground

Fomitiporia fissurata　　　　　*Pseudoinonotus dryadeus* group

F. fissurata is a perennial that can reach > 12 in. diameter. *P. dryadeus* is an annual known as the giant oak polypore and can be 12–36 in. wide; may possibly be two species. Only one polypore is larger—*Bridgeoporus nobilissimus* (image on page 6), which is a protected species where it grows. Its DNA can easily be found on many tree species. *B. nobilissimus* is deeply fuzzy, often with mosses and small trees growing from its top. It is found on old-growth noble firs. Never harvest one.

91d. Bristly semicircular bracket with rough, reddish brown pore surface

Inonotus hispidus

Normally on hardwood, especially oaks, in OR and CA. Can be ≤ 6 in. wide, 4 in. long, and 3 in. thick. Large, angular pores with torn walls may exude sulfur-yellow droplets. Usually without a stipe. Pleasant odor, mild to astringent taste. Inedible. Global distribution. (See also *Funalia 'gallica'* IN01 [page 316] on cottonwoods.)

92a. (84a) Flat on substrate, not forming any cap *94a*

92b. May form a small cap (see also key leads 76a–d [pages 100–101]) *93a*

93a. (92b) Fruitbody gelatinous, ≤ 10 in. wide, 4 in. tall, and ¼ in. thick

Phlebia tremellosa

This distinctive species was (and will likely again be) known as *Merulius tremellosus*. All parts are gelatinous. Mainly found on hardwoods, sometimes conifers, from July through January. May remain flat on the surface of the wood or form distinct caps. No odor or taste. Inedible. Global distribution.

↓93b. Crust and cap white to ochre, pores tooth-like

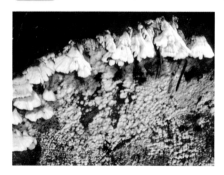

Irpex lacteus

Annual on dead hardwoods, sometimes conifers, ≤ ½ in. tall, 3 in. wide, and ¼ in. thick. Pore surface soon becomes tooth-like. White to ochre caps may or may not form along the upper edge of the crust. Faded *Trichaptum* species can be very similar. Inedible. Distribution NA, EU.

94a. (92a) Polypore growing flat on wood, adhering tightly

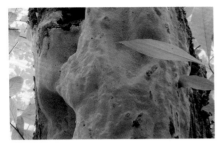

Perenniporia subacida

Pore surface ivory-yellowish to tan. Found year-round, mainly on conifers in NA. The specimen shown at left covered about 60 ft. of a large conifer. Distinctive. Taste is slightly sour. Tough, fibrous, and inedible.

↓94b. Annual polypore grows flat on dead hardwoods and conifers

Fuscoporia ferruginosa   *Fuscoporia ferrea*

F. ferruginosa (NA, EU, Asia) is tough, adheres strongly to the wood, and is ≤ 1 in. thick. It is rusty reddish brown. *F. ferrea* (NA) is yellow-brown. The odor and taste of both is indistinct.

↓94c. Loose white material densely covered in narrow white to brownish spines

Hydnocristella himantia
(= *Kavinia himantia*)

Whether the "correct" genus should be *Hydnocristella* or *Kavinia* is unresolved. It grows on both very decayed hardwoods and conifers and can spread to adjacent loose debris and soil. I have seen it only in the fall mushroom season and presume it is an annual. Distribution NA, EU.

↓94d. Fruitbody white, then bright orange, rose, or red; usually on an old polypore

Hypomyces aurantius

H. rosellus (see key lead 3a #4 [page 44]) starts white and soon turns rose and then reddish. *H. aurantius* soon turns orange and then reddish. Both grow on old polypores and sometimes gilled mushrooms and can spread out over adjacent branches, sticks, and vegetation. Inedible. Global distribution.

↓94e. Fruitbody white to pale ochre, soft, pore-like

Trechispora mollusca

Soft, fragile, annual species grows on polypores, hardwoods, and conifers. Taste is mild. Large open pores are distinctive. Inedible. Global distribution.

↓94f. On hardwood; fruitbody black (at least at maturity) *96a*

94g. On hardwood, rarely conifers; fruitbody white to orangish *95a*

95a. (94g) Whitish covering on wood; tiny spines on underside when mature

Lyomyces crustosus
(= Hyphodontia crustosa)

Typically found on hardwood branches. Coating has a waxy look and develops short spines when mature. Found year-round. Several similar *Hyphodontia* species can be differentiated only by microscopic features. Inedible. Distribution NA, EU, Asia.

↓95b. Yellowish to orange fruitbodies, white pruinose border, ± confluent

Peniophora eriksonii

Found on hanging alder branches late summer through fall. The similar *P. aurantiaca* (p. 316) can be distinguished only by microscopic features. Both in NA, EU.

95c. Small, strongly cracked tubercles on oak

Xylobolus frustulatus

Distinctive species is a perennial that can be found year-round. Only known from oak, where it forms a crust, ≤ ½ in. thick, that soon cracks into small polygon shapes. Hard and inedible. Distribution NA, Mexico, EU.

96a. (94f) Young growth soft, grayish white, soon black and brittle

Kretzschmaria deusta

Grows on hardwoods, most frequently bigleaf maple. Typically found at the base of old stumps. Very distinctive (an ascomycete), forming a < ¼ in. thick crust. Inedible. Found globally. (See also *Rhizina undulata* [page 316].)

96b. Dark brown to black at all growth stages

Biscogniauxia mediterranea

Grows on oak branches and other hardwoods after drought stress or fires. Flat, black cushions are dotted with openings for spore release. If the fruitbody is more globular and completely covered with tiny mounds, check out *Annulohypoxylon thouarsianum* and *Daldinia grandis* in MycoMatch. All three are globally distributed ascomycete fungi.

97a. (2f) Found ± above ground; distorted gills (page 123) *105*

↓97b. Resembles a thick membrane crumpled into a ball (page 122) *104a*

↓97c. With a thin skin *99a*

97d. With a thick skin; interior powdery black at maturity *98a*

98a. (97d) With a distinctive rounded columella (vestigial stipe)

Schenella simplex

Fruitbody diameter 1–2 in. Interior spore mass goes from white to gooey and blackish to powdery black. The rounded columella is distinctive. Usually found near conifers and along old roads. Unpleasant odor in age. Inedible. Distribution western NA into Mexico.

98b. Columella absent; diameter < 2 in.

Elaphomyces species

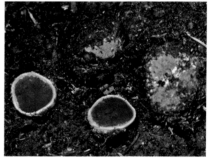

Elaphomyces asperulus   *Elaphomyces 'muricatus'*

E. asperulus has long been known under the EU name *E. granulatus* and is the most common Cascadia species of several hard-to-distinguish species in a complex. The thick skin of *E. asperulus* is reddish gray to sordid red–violaceous. Common year-round in Cascadia under conifers, oaks, and chinquapin. Widespread in the Northern Hemisphere. *E. muricatus* is a EU species that has not been found in Cascadia (except for DNA in a soil sample) but has several look-alikes here. The known western species in the *E. muricatus* complex are yet to be named. All species start hollow but soon fill with cottony material and then powdery, dark brown to black spores. None are edible.

99a.   (97c) Interior lacking a columella (vestigial stipe) (page 120) *101a*

99b.   Interior with a columella *100a*

100a. (99a) Yellowish green with a short, thick columella; 1–3 in. wide and 1–2 in. tall

Truncocolumella citrina

Related to the bolete genus *Suillus*, this species is common under Douglas fir and sometimes other conifers. Grows just under the duff and is often exposed. Very distinctive, but tasteless until mature and turns gooey. Edible. Western NA.

↓100b. Tan with a thin, indistinct columella

Gautieria monticola

One of more than twenty *Gautieria* species, all characterized by a rubbery interior filled with irregular spore-bearing cavities. ≤ 4 in. wide. Under conifers, often exposed, mainly above 3,000 ft. but also down to sea level. Pleasant smelling collections are edible but rubbery. Western NA.

100c.   Whitish; leathery skin separates from gleba; diameter < 1 in.

Hysterangium coriaceum   Hysterangium crassirhachis

Half a dozen *Hysterangium* species have been found under both hardwoods and conifers. The columella can be indistinct. None are edible. Western NA.

`101a.` (99a) Interior looks and feels like a dense rubber ball, bounces (page 122) *103*

`101b.` Interior irregular chambers separated by differently colored veins *102a*

`102a.` (101b) Under alders; reddish bruising; diameter < 1½ in.

Alpova species

Alpova diplophloeus   *Alpova concolor*

Alpova species, known as red gravel, are common under alders. *A. diplophloeus* has a single-layered skin. *A. concolor* has a double-layered skin. Interior of both is sticky-gelatinous and may exude juice. Grows in late spring through early winter. Odor is fruity at maturity. Edibility? Found in NA.

`↓102b.` Under the coastal strain of Douglas fir; diameter ≤ 2 in.

Tuber oregonense   *Tuber gibbosum*

Four choice fungi known as Oregon white truffles are worth more than $800 per pound when mature and smell cheesy-garlicy, but they're worthless if not ripe. Oregon white, Oregon black, and Oregon brown truffles are found with the coastal strain of Douglas fir (as far east as White Salmon, WA, north to southern BC, and south into northern CA). Found below about 2,000 ft. in densely planted second growth trees that are 15–40 or more years of age. Hunt with a truffle dog.

↓102c. Mainly under oaks, sometimes conifers

Tuber candidum complex

I sometimes uncover these < 2 in. diameter truffles in April to June under oak trees. Twice I have tried to find mature ones using a friend's truffle dog. Immature truffles have little odor. Edible, tasteless unless ripe. Transcontinental, southern Canada to Mexico. *T. luomae* (page 316, inedible) grows under Douglas fir WA.

102d. Under hardwoods and conifers, blackish pockets separated by whitish to yellowish veins

Melanogaster species

Melanogaster euryspermus   *Melanogaster 'tuberiformis'*

Several *Melanogaster* species (diameter ≤ 2 in.) grow in Cascadia. *M. euryspermus* has blackish gel-filled chambers separated by whitish to yellow veins. *M. tuberiformis* (Cascadia, EU) has whitish veins and firm, black pockets that turn gelatinous in age. Both are typically unpleasant smelling when ripe, thus are undesirable as edibles. Both ripen in March and April under hardwoods and conifers. One unidentified species with a pleasant odor had an interesting earthy licorice flavor. The value of truffles is all in the odor, and sometimes *Melanogaster* species smell great and are delicious. None are toxic. The exterior of *M. euryspermus* somewhat resembles both the Oregon brown truffle, *Kalapuya brunnea*, and the Oregon black truffle, *Leucangium carthusianum*. Both mainly fruit in fall to winter under coastal Douglas fir with other conifers and hardwoods. *K. brunnea* has an interior with whitish veins and grayish brown pockets and is mildly cheesy-garlicky at maturity. *L. carthusianum* resembles animal dung; the interior has white veins with grayish green pockets; odor ± like pineapple.

103. (101a) Fruitbody minutely chambered

Rhizopogon vulgaris (young at left, mature at right)

Rhizopogon ellenae   *Rhizopogon villosulus*

It is incredibly easy to determine that you have found a species of *Rhizopogon*. Drop it on a hard surface and it will bounce. Figuring out which species you have found is tough, requiring chemical reagents, microscopy, and usually DNA testing. More than 125 *Rhizopogon* species are found in in Cascadia, year-round under conifers. Diameters 1–4 in. None are known to be toxic; few are worth eating. All are in the Boletales. Probably globally distributed, anywhere *Suillus* species occur.

104a. (97b) Hollow or a folded interior; exterior not fuzzy

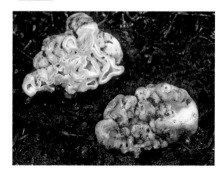

Hydnotrya variiformis
(= *Gyromitra variiformis*)

Minutely velvety on the exterior. Found under varied conifers from May through October. Diameter ≤ 1½ in. Up to six species in Cascadia. Lumpers (my favorite people) put these in genus *Gyromitra*. Not common. No odor or taste. Edibility? Western NA.

104b. Highly convoluted, fuzzy brown, soil-encrusted outer surface

Geopora cooperi

Though *G. cooperi* are typically golf ball–sized, I once encountered numerous baseball-sized specimens in a burn area. Usually found under conifers, sometimes hardwoods, spring through fall (western NA, EU). Mild odor. Reportedly well worth eating.

105. (97a) Arid habitat species with or without a well-developed stipe

Battarrea phalloides group (both)   *Tulostoma* cf. *americanum*

Montagnea arenaria   *Chlorophyllum agaricoides*

The distinctive *C. agaricoides* is found in lawns, fields, and waste spaces in spring to fall. Edible when young and can reach 4 in. wide and tall, while *Agaricus inaperatus* (page 316) is smooth and inedible. None of the other species here are edible. Members of the *B. phalloides* group develop underground in a volva, and the stipe can be ≤ 16 in. long with a 2 in. spore case. There are > thirty species of *Tulostoma* in Cascadia alone; most are 1–2 in. tall and 1 in. wide. *M. arenaria* is a distinctive gastroid *Coprinus* species, ≤ 10 in. tall and ≤ 1½ in. wide. These species occur in arid areas worldwide.

106a. (2e) Nest with tiny "eggs" (bird's nest fungi) (page 131) *116a*

↓106b. Thick outer layer splits into rays and folds back under the fruitbody, raising it to surface; inner layer disintegrates to release powdery dark spores (earthstars) (page 130) *115a*

↓106c. Spore mass (gleba) white when young, with look and softness of a marshmallow (puffball) (page 126) *108a*

↓106d. Spore mass white when young but very firm (earthballs) *107a*

↓106e. Fruitbody with an exposed upper part that develops a powdery brown spore mass; hard black and yellow below ground portion

Pisolithus 'arhizus' group

Long known as *P. tinctorius,* now a synonym of *P. arhizus* (EU), this large, grotesque mushroom (≤ 8 in. wide and 12 in. tall) is a worldwide complex of three look-alike species. It is prized by dye makers and abundant in dry oak-pine woods. Found July through October in the Northern Hemisphere.

106f. With a stipe; spore mass on surface of cap stinks of rotten flesh

Lysurus cruciatus

Mutinus ravenelii   *Phallus hadriani*

These three stinkhorn species emerge from an egg-like volva and are technically edible but not desirable. The surface of the head is covered with a stinking slime layer that attracts flies to disperse the spores. The volva (egg) is white in *L. cruciatus* and *M. ravenelii* but pinkish purple in *P. hadriani* (some consider this a variant of *P. impudicus,* which has a white volva). All three are uncommon in the Pacific Northwest though reported to grow in a wide variety of habitats. Exactly which species are in Cascadia is unclear since we have no DNA data. Globally there are eighty species of stinkhorns (mostly in warm regions) and many have elaborate shapes.

107a. (106d) Smooth, cracking in age; spore mass bruises red when young; ½–2 in. diameter

Scleroderma 'areolatum'

Stem-like base with well-developed mycelial rhizomorphs. Spore mass turns powdery and dark violet to olive-brown with age. Odor and taste slightly sweet. Causes rapid gastric distress (and possibly near-death) if consumed and can kill dogs and pigs if they eat it. Global distribution. The Cascadia species is distinct.

↓107b. Exterior with distinctive scales in a rosette pattern; ≤ 4 in. wide

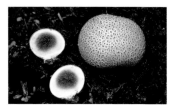

Scleroderma 'citrinum'

Surface bruises a faint dark red when young. Interior white, then violet, and finally dark olive-brown. Unlike S. areolatum, S. citrinum is almost stemless. Odor and taste are mild to bread-like. Rapidly causes gastric distress, chills, and sweats if consumed. Can kill dogs and pigs if they eat it. Global distribution. The Cascadia species is distinct.

↓107c. Rhizomorphs at base ± 70 percent as wide as fruitbody; ≤ 3 in. wide

Scleroderma 'meridionale'

At maturity, the top of the fruitbody opens in a star-like manner, and initially white gleba turns purple-brown and becomes powdery. Grows in sand near hardwoods and conifers. Little odor or taste. Edibility? Considered poisonous. NA, EU, Russia. The Cascadia species is distinct.

107d. Develops underground in poor soil and opens star-like; no stipe; ≤ 5 in. wide

Scleroderma 'polyrhizum'

Interior whitish with yellow filaments when young, powdery purplish black at maturity. On waste ground and pastures, rarely in woods. Odor and taste not reported. Poisonous. Global distribution. The Cascadia species is distinct.

108a. (106c) Fruitbody typically ± golf-ball size *110a*

108b. Fruitbody typically > baseball size *109a*

109a. (108b) Alpine species; long pyramidal warts; fruitbody 3–6 in.

'Calvatia' sculpta

Warts on cap can be > 1 in. long and are often coiled. The mature spore mass is dark olive-brown. A very distinctive edible species (only when all white inside). Found late spring through fall under conifers. Most common in CA, also in OR and WA. Belongs in genus *Lycoperdon*.

↓109b. Above 2,000 ft., pineapple-like warts

'Calbovista' subsculpta

Can reach basketball size. Found as early as April at > 2,000 ft. and continuing into fall as high as the tree line. Interior turns brown to purple-brown at maturity and the pineapple-like scales fall off, exposing the spore mass. Edible if pure white inside. Western NA mountains. Belongs in genus *Lycoperdon.*

↓109c. Grassland species with purple spores

Calvatia fragilis

Formerly known as *C. cyathiformis.* Exterior brownish to light purple, cracks like mud at maturity. Can be ≤ 4 in. wide and 2 in. tall. Distinctive purple spore mass at maturity. Found in the warmer, dryer inland areas in grass and bare soil as soon as the fall rains arrive. Distinctive. Edible while still pure white inside. Distributed globally.

↓109d. In pastures and barnyards in both humid and arid areas

Mycenastrum corium

Felty outer layer cracks like mud and wears away to reveal thin, tough inner membrane that wears away to reveal dark brown to purple-brown spore mass. Mild to pleasant odor and taste. Edible when firm and white inside. Global distribution.

109e. Near sagebrush or grass in arid areas; diameter ≤ 2 ft.

Calvatia booniana

Cottony exterior breaks into polygonal warts, disintegrating at maturity to reveal a greenish to brownish gleba. Odor mild to unpleasant. Edible when firm and pure white inside. Consume with caution—it can have a laxative effect. Distribution western NA.

110a. (108a) Skin single or double layered, but thin; in both alpine and low-elevation areas (page 128) *112a*

110b. Skin thick; subalpine to alpine species *111a*

111a. (110b) Surface smooth, sometimes finely cracked like mud

'Gastropila' fumosa

Common. Formerly known as *Calvatia fumosa*. Diameter 1–4 in. Found next to melting snowbanks at high elevation, persisting into summer. Surface white to brownish, mature spore mass dark brown. Odor unpleasant, like flatulence. Inedible. Belongs in genus *Lycoperdon*. *Lycoperdon 'vernimontanum'* (page 317) is similar. Distribution western NA.

111b. White surface breaks into dark-tipped, polygonal warts

Lycoperdon subcretaceum (= *Gastropila subcretacea*)

Formerly known as *Calvatia subcretacea*, this species appears shortly after snowmelt above 3,000 ft. It can reach baseball size and is edible when young and pure white inside. Spore mass olive-brown at maturity. Never eat any maturing puffball that has started to change color. Distribution western NA mountains.

112a. (110a) Lacking tiny, pointed warts that readily break off *113a*

112b. With tiny, pointed warts that readily break off

Lycoperdon perlatum

When spines are rubbed off, distinctive polygonal scar remains on the white surface. Found at all elevations in the woods, on lawns, and on cultivated ground. Common, < 3 in. wide, often found in sufficient quantity to pick for eating. Puffballs are edible only when the interior is white. Global distribution.

113a. (112a) Outer skin layer separates from inner layer *114a*

↓113b. Outer layer not separating; on lawns, distinct sterile base

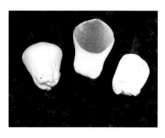

Vascellum lloydianum/V. pratense

Typically, ± top-shaped, 2 in. tall and ≤ 2 in. wide. Exterior with soft spines that fall away at maturity. Exterior initially white, aging dark metallic brownish. Opening via a wide tear at top. Odor, taste, and edibility are not reported. Two sister species? Distribution NA, EU, Asia.

↓113c. Outer layer not separating; with dark brown to black spines; 2 in. tall and ≤ 1½ in. wide

Lycoperdon nigrescens

Spines fall off in age, leaving a netted pattern on the inner spore case. Opens by a pore at the top. On debris in conifer or hardwood forests, summer to fall. Odor unpleasant, taste mild. Edibility? Distribution NA, EU.

113d. Outer layer not separating; ± persistent granular coating

Lycoperdon umbrinum

Like *L. nigrescens*, with tiny, pointed warts that do not easily rub off and do not leave a polygonal scar; visually separated by color of the skin. *L. umbrinum* is < 3 in. wide. Found at all elevations in coniferous and mixed woods, often on wood, summer to fall. Edibility? Distribution NA, EU, Asia.

114a. (113a) Smooth outer skin peels off like an eggshell

Bovista plumbea   *Bovista pila*

Both species found widespread in meadows and pastures in spring through fall. At maturity, the spore case breaks loose and is blown by the wind. *B. pila* (in Cascadia) has a metallic bronze inner skin and can be 4 in. wide. *B. plumbea* (NA, EU, Russia) is ≤ 2 in. wide and has a metallic blue-gray to lead-colored inner skin. Neither is worth eating.

114b. Outer skin with distinct spines that flake off in sheets; ± 2 in. wide

Lycoperdon marginatum

Outer skin is white when young, aging light to deep brown. Spore mass olive to grayish brown. Found June to October in oak-pine woods on poor soil. Mild taste, edible. Global distribution.

115a. (106b) At maturity, spores are released via small pore at the top

Geastrum saccatum   *Geastrum triplex*

Eighteen known *Geastrum* species are in the Pacific Northwest. All open via a pore at the top to release the spores. In some, the inner spore case is raised above the rays by a short column. In these two species, the spore case sits flat on the rays. Both grow in mixed woods, mostly with hardwoods present. In *G. saccatum* (± 2 in. wide), spores are purplish brown, but *G. triplex* (opens to 5 in.) spores have no hint of purple. *G. triplex* always has rays that split to give a saucer-like base under the spore case. *G. triplex* is highly variable. Both are too tough to eat. Both globally distributed.

115b. At maturity, spores released via a torn pore, rays close in dry weather

Astraeus hygrometricus group   *Astraeus pteridis*

Either *A. hygrometricus* (opens to 3 in.) or a highly similar, newly named species, *A smithii*, can be found from sea level up to tree line in open and waste areas and persists year-round in Cascadia. *A. hygrometricus* or its look-alikes are distributed globally. *A. pteridis* (opens to 6 in.) is distinguished by the distinct checking on the rays. It is found mostly in open waste areas in Cascadia. Both are inedible.

116a. (106a) Single tiny, dark "egg" (periodole) in a minute, star-shaped cup

Sphaerobolus stellatus
(± 2x life-size)

This ⅛ in. diameter fungus contains a single, dark periodole that is shot out like a cannonball by the sudden inflation of a whitish sack under the spherical periodole (see top of image). Grows on plant matter, wood, or manure, and loves bark mulch. *S. stellatus* and the look-alike *S. iowensis* are known from NA, EU, Africa.

116b. With multiple periodoles in a ¼–½ in. splash cup *117a*

117a. (116b) Periodoles held in nest by a thin cord

Crucibulum   *Cyathus stercoreus*
'crucibuliforme'

Both cups start covered by a whitish, coarsely hairy lid (epiphragm), and as the spores mature, the lid disintegrates. In *C. stercoreus* the mature periodoles are dark brown to black; they are whiteish in *C. 'crucubuliforme'*
(= *C. laeve*). Raindrops landing in the cup splash the periodoles a considerable distance. The sticky cord catches on vegetation. Empty cups seen year-round. Global distribution.

117b. Periodoles held in cup by gelatinous matrix, no-cord *Nidula* species

Nidula   *Nidula candida*
niveotomentosa

Globally distributed *N. niveotomentosa* (¼ in. tall and wide) has mahogany-brown periodoles while Cascadia species *N. candida* (⅓ in. tall and 1 in. wide) has pallid to gray-brown periodoles. Both grow on sticks, bracken, and berry canes (plus other vegetation for *N. candida*). Cups persist year-round. Inedible.

Note: When identifying brightly colored *Ramaria* species, at least one specimen must be young, and colors must be noted the day the fungus is picked. Also, the specimen must be harvested with its entire base. Once home, cut the specimen in half from top to bottom to help identify to species. Note flesh texture (gelatinous, cartilaginous, or fleshy) and look for a rusty stain at the interior base of the stipe.

118a. (2d) Branches or tips pink, red, or purple red (page 143) *137a*

↓118b. Branches and branch tips bright orange (page 142) *133a*

↓118c. Branches orange or pink, tips yellow (page 140) *129a*

↓118d. Branches and tips yellow or orange (page 138) *126a*

118e. Branches various, none of the above (118a—118d) *119a*

119a. (118e) Branches and base fleshy (page 136) *123a*

↓119b. Branches and base very slender *120a*

↓119c. Branches flattened; tips broad and flattened; leathery texture

Thelephora 'palmata'   *Thelephora terrestris*

Globally distributed *T. terrestris* is reddish brown to smoky gray-brown, with a paler fringed margin; 1–2 in. wide, or rosettes of 5 in. wide and 3 in. tall. Odor like moldy earth. *T. palmata* (NA, EU, Asia) is profusely branched with purplish brown to chocolate-brown branches and paler tips; 1–4 in. tall and ≥ 4 in. wide. The NA species may be distinct. Odor strong and garlic-like. A dozen *Thelephora* species reported from Cascadia.

119d. Branches flattened; tips broad and flattened; fleshy

Sparassis radicata

Distinctive. Can grow as large as a bushel basket. The odor is ± strong and spicy-fragrant. Exceptional edible. When harvesting, cut off at ground level, leaving the "root" in place. *S. radicata* (Cascadia) is a sister species to *S. crispa* (EU). *S. spathulata* (eastern NA) has larger lobes.

120a. (119b) Growing on rotten wood or from buried wood (page 135) *122a*

120b. Growing on the ground *121a*

121a. (120b) Stipe covered with chestnut-colored, short, bristly hairs

Clavulina castaneipes var. *lignicola*

Fruitbody 1–3 in. tall. Branches flattened, rose-pink to dull tan or lead-gray. Grows on bark or ligneous duff and is slightly bitter and inedible. Uncommon, season not well defined. Global distribution of sister species, though *C. castaneipes* var. *lignicola* itself appears to be a unique western NA species.

↓121b. Base short (< 1 in.) or absent; whitish or grayish crested branch tips

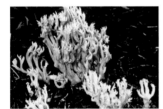 

Clavulina 'coralloides'   *Clavulina cinerea* group

C. 'coralloides' (< 2 in. wide and < 3 in. tall) is characterized by crested branch tips and was formerly known as *C. cristata*. The DNA of *C. coralloides* (EU) is distinct from the western NA species, which needs a new name. The *C. cinerea* group (< 5 in. wide and tall, north temperate regions) is similar but with smoky gray on the branches. When parasitized with *H. clavariarum*, either species turns a beautiful, deep blue-black. *C. 'coralloides'* and *C. cinerea* can be found from early summer through late winter under both hardwoods and conifers. Both have little odor or taste but can be eaten if not parasitized. There are six somewhat similar ± inedible *Phaeoclavulina* species in Cascadia.

Helminthosphaeria clavariarum

↓121c. Covered at base with a mat of white mycelium; not bruising

Phaeoclavulina curta

Fruitbody < ¾ in. wide and 1½ in. tall. Yellow-ochre branches, with lighter, often forked tips, EU, NA (probable). Odor and taste not recorded. *Ramaria myceliosa* var. *microspora*; with small, but slightly larger spores than *P. curta*, has been considered identical but is now separate. *R. myceliosa* var. *myceliosa* has even larger spores and a mild to aromatic odor and a bitter taste. Inedible. All three species/varieties are in EU and at least sister species are in NA.

↓121d. Covered at base with a mat of white mycelium; bruising blue-green

Phaeoclavulina abietina group

Formerly *Ramaria abietina*. Fruitbody < 1½ in. wide and < 3 in. tall. Variable stipe, often slender or indistinct. Odor can be earthy or anise-like. Taste is bitter. A complex of several species. *P. mutabilis* slowly stains reddish brown. All group members and *P. mutabilis* are inedible. Distribution NA, EU.

121e. White branches, rarely tinged flesh-color; no mycelial mat; no bruising

Ramariopsis 'kunzei'

Fruitbody ≤ 3 in. wide and 4 in. tall. Branch tips are not crested. Somewhat fragile. On ground or sometimes woody debris under conifers, in mixed woods, and in fields. Little odor or taste, nontoxic. Two distinct sister species in Cascadia. Global distribution.

122a. (120a) Branch tips with multiple crown-like ends

Artomyces piperatus

Formerly *Clavicorona piperata*.
Fruitbody 1–3 in. tall. Found on
rotten conifer logs in the fall west
of the Rockies. Odor indistinct.
Peppery taste. Too tough to eat.
A. cristatus (northern NA, EU) has
no reported odor or taste and
differs microscopically.

↓122b. Branch tips pointed or toothed; from tangle of rhizomorphs

Lentaria pinicola

Fruitbody < 3 in. tall. Multiple
branches, ± a stipe, from a tangle
of mycelial cords. On rotting
conifers. Odor earthy, taste slightly
bitter, inedible. *L. epichnoa* (more
slender, white) differs enough
to probably belong in a different
genus. *L. pinicola* may also be moved to a new genus. Distribution *L. pinicola*
(Cascadia), *L. epichnoa* (also in EU).

122c. Copiously branched; tips forked; from tangle of rhizomorphs

Ramaria rubella   *Ramaria stricta*

These < 4 in. tall species grow mainly on rotting conifer wood, summer and fall in
NA and EU. Both are peppery and inedible.

123a. (119a) Flesh not discoloring; no "rusty root"; not gelatinous *124a*

↓123b. Gelatinous streaks in flesh; fruitbody < 6 in. tall and 5 in. wide

Ramaria gelatinosa var. *oregonense*

Light orange-brown and ages to grayish–light brown. Widespread under conifers, especially hemlocks, in late summer through fall in Cascadia. Easy to recognize if cut in half. Musty-sweet odor. Reportedly causes GI distress for some after eating.

↓123c. Cut flesh rapidly bruises red-brown; fruitbody < 6 in. tall and 4 in. wide

Ramaria testaceoflava var. *brunnea*

Instantly recognizable from the bruising reaction. Found late summer through fall under western hemlock. Fairly common. The odor is faint to earthy, taste mildly bitter. Not a good edible. Variety *brunnea* may be unique to western NA. Distribution NA, EU.

123d. Rusty region at very base of stipe interior (not a bruising reaction); large

Ramaria velocimutans

Fruitbody < 10 in. wide and 12 in. tall. This species is distinguished by a rapid blue-green reaction on contact with iron salts. Found late summer through fall under hemlocks and yew in Cascadia. Common. Odor sweet. No poisonings known.

124a. (123a) Fruits in the spring season *125a*

↓124b. Fruits fall; taller (< 12 in.) than wide (< 7 in.)

Ramaria acrisiccescens

Branches are pale yellowish brown to pale orangish brown, tips are pallid to pinkish, and lower branches bruise brown. A very undistinctive coral. It fruits in the fall under western hemlock in Cascadia. The odor is musty and green bean–like and the taste is bitterish acid. Inedible.

124c. Fruits fall; ± as wide (< 4 in.) as tall (< 5 in.)

Ramaria caulifloriformis

Pale buff to pinkish buff branches; cartilaginous to slightly gelatinous flesh in stipe. Flesh colored much like surface. Odor and taste mild to green bean–like. Fruits under western hemlocks and western red cedar in Cascadia and MI. Edible. If the branches have a violet tinge, see *R. subviolacea* (page 317).

125a. (124a) Branches light brown to fleshy tan; fruitbody < 4 in. wide and 6 in. tall

Ramaria marrii

The single whitish base bruises brown. Fairly distinctive. Fruits under white pine and grand fir throughout Cascadia. Odor is indistinct, and taste is slightly bitter and mildly astringent, making it a poor choice for the table.

125b. Branches off-white; fruitbody < 4 in. wide and 5 in. tall

Ramaria rubricarnata var. *pallida*

Branches are off-white to pale salmon-pink. Branch tips are pale greenish yellow, soon pale yellow. An uncommon species found under mixed conifers in Cascadia. The odor and taste are mild to slightly green bean–like. Edible.

126a. (118d) Fruiting in the spring, sometimes summer to fall *128a*

126b. Found in fall season (only two of several species shown) *127a*

127a. (126b) Citrus-like odor; yellow branches and tips; white basal tomentum

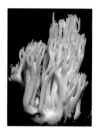

Ramaria cystidiophora var. *citronella*

Fruitbody < 5 in. wide and 7 in. tall. Common fall species is clear and bright yellow when young. Under conifers, all elevations in Cascadia. Identified by the lemony odor. Taste is slightly bitter. *R. cystidiophora* var. *fabiolens* differs in having a green bean–like odor and tends to be taller and broader. *R. cystidiophora* var. *maculans* (page 317) stains reddish brown.

127b. Vinaceous stained base; fruitbody < 7 in. tall and wide

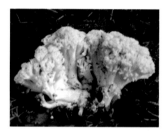

Ramaria rubiginosa

Young branches light yellow with bright yellow tips. Under western hemlock in Cascadia. No distinctive odor or taste. No reports of toxicity. *R. vinosimaculans* (see 128e), a spring species, also stains reddish brown and can occur in the fall. *R. cystidiophora* var. *maculans* (page 317) is another red-brown–bruising species.

128a. (126a) Fruitbody (< 8 in. tall by 10 in. wide) cream to yellow; tips frost red

Ramaria rasilispora var. *scatesiana*

Under true firs in spring. Edible. Hard to distinguish from slightly darker *R. rasilispora* var. *rasilispora* and *R. magnipes* group. Distribution Cascadia, possibly Finland and Poland.

↓128b. Fruitbody (≤ 7 in. wide and 6 in. tall), yellow-orange to orange

Ramaria magnipes

Several yellow to light orange spring *Ramaria* species with a large fleshy base can grow in stupendous abundance in the spring (often in fairy rings) in Cascadia. Can be difficult to differentiate, even with a microscope (see also *R. magnipes* PNW01, a sister species, page 317). All are bland and edible.

↓128c. Branches and tips pastel orange to apricot; massive stipe

Ramaria armeniaca

Huge base gives an ovoid shape, 4 in. wide and 5 in. tall. Branches orange above, white below. No odor and mild taste. Edibility? Cascadia species.

↓128d. Yellow-orange branches, tips yellow; yellow band at soil level

Ramaria rubricarnata var. verna

Fruitbody < 3 in.wide and 4 in. tall, upper branches salmon-orange to light salmon. Solid white flesh, odor ± like green beans, no taste. Edibility and habitat not well established. Uncommon. Specimen shown was found at 3,000 ft. under mixed conifers in Cascadia.

↓128e. With reddish to violet stains at the base of the branches

Ramaria vinosimaculans

Fruitbody ≤ 8 in. tall and 8 in. wide. The mushroom does not bruise but is found with stains on the branch base. Grows under a wide range of conifers in both spring and fall in Cascadia. Odor is somewhat sweet and taste is mild. Not very abundant, but a decent edible.

128f. Pallid when covered by duff, yellow in sunlight; brown bruising

Ramaria thiersii

Fruitbody < 3 in. wide and 6 in. tall. The bruising reaction ranges from weak to strong. Found under true firs and mixed conifers in June above 2,500 ft. in Cascadia. Not abundant and not with a large, meaty base, so not tested for edibility. Odor and taste are mild.

129a. (118c) No "rusty root" when cut vertically through the center *131a*

129b. With a "rusty root" *130a*

130a. (129b) Base surface smooth, branches peach- or salmon-colored

Ramaria amyloidea (above and at right)

Fruitbody < 6 in. wide and 6 in. tall. Flesh soft and stringy, taste mild, and odor slightly sweet. In the fall under hemlock or fir in Cascadia (considerable genetic variation may indicate a group). Distinguished from all but *R. celerivirescens* and *R. clavariamaculata* because of the rusty root. Common, but not abundant. Edibility?

130b. Base surface rough, branches yellow-orange

Ramaria celerivirescens/R. clavariamaculata, young (at left) and old (at right)

Two look-alikes, fruitbody < 4 in. wide and 7 in. tall. Branches pale to light orange to light pinkish orange; tips light to sunflower-yellow. Found in fall under western hemlock in Cascadia. Taste is mild to bitter and odor unpleasant. Edibility?

131a. (129a) Base and/or lower branches not bruising

Includes several fall *Ramaria* species that are difficult to separate without a microscope and that I have not reliably photographed. These include *R. rubricarnata* var. *rubricarnata*, *R. cyaneigranosa* var. *persicina*, *R. longispora*, and *R. leptoformosa*.

↓131b. Base flesh subgelatinous to gelatinous; dull violet bruising *132a*

131c. Base flesh fleshy and stringy; bruising brown when handled

Ramaria formosa

Fruitbody < 4 in. wide and 6 in. tall. Young branch tips light yellow; branches light reddish orange; base white, bruising brown. Under Douglas fir, hemlock. Odor and taste mild. No poisoning reports in NAMA database. Western NA has a distinct variety. Distribution NA, EU.

132a. (131b) Base a bundle of stringy, white stipes; < 1½ in. wide and 4 in. tall

Ramaria conjunctipes var. *sparsiramosa*

Only differs from *R. conjunctipes* var. *tsugensis* by its smaller stature and thinner branches. Dull violet bruised spots. Texture fleshy-pliable or rubbery. Under western hemlock in the fall in Cascadia. Odor and taste mild. Edibility?

132b. Base single to clustered, stipes not stringy, < 3 in. wide and 7 in. tall

Ramaria conjunctipes var. *tsugensis*

When young, branch tips are yellow and branches are pale to light pinkish orange with a waxy-translucent look. Edibility? There are numerous other somewhat gelatinous, dull purple–bruising species in Cascadia, including *R. sandaracina* var. *euosma*, *R. gelatiniaurantia* var. *violeitingens*, *R. flavigelatinosa* var. *megalospora* (page 317), and *R. flavigelatinosa* var. *carnisalmonea*.

133a. (118b) "Rusty root" present (see 130a and 130b [page 140])

133b. No "rusty root" *134a*

134a. (133b) Base flesh lacking gelatinous streaks when cut vertically *136a*

134b. Base flesh showing gelatinous streaks when cut vertically *135a*

135a. (134b) Odor musty-sweet; base very gelatinous (see 123b [page 136])

135b. Branches bright orange with a yellow band just above the white base

Ramaria sandaracina var. *chondrobasis*

Fruitbody < 5 in. wide and 6 in. tall. Branch tip flesh ± branch color. Sometimes dull violet bruised areas. Under western hemlock in Cascadia. Odor ± slightly like green beans. Edibility? *R. gelatiniaurantia* var. *gelatiniaurantia* has yellow flesh in branch tips, green bean-like odor.

136a. (134a) Lacking a yellow belly band when young; not bruising

Ramaria largentii

Fruitbody < 6 in. tall and wide. Base flesh soft and stringy. Odor slightly sweet, taste mild. In fall under lowland hemlocks in western NA, EU. *R. flavigelatinosa* var. *megalospora* (page 317) is distinguished by a ± beany odor and mild taste with base flesh-like cartilage. In the fall under hemlocks in NA.

136b. Yellow belly band when young and/or bruising brown or violet

Ramaria sandaracina var. *sandaracina*

Fruitbody < 3 in. wide and 6 in. tall. Distinguished by brittle, fleshy-fibrous, slightly gelatinous flesh. Under western hemlock and Douglas fir in the fall in Cascadia. No poisoning reports. *R. aurantiisiccescens* (page 317) has a sweetish odor and no brown staining.

137a. (118a) Branch tips wine-red to purplish red (page 144) *139a*

137b. Branches pink, orange-red, to crimson-red, tips red to yellow *138a*

138a. (137b) Branches ± bright peachy pink

Ramaria cyaneigranosa varieties
Ramaria cyaneigranosa var. *cyaneigranosa*

Fruitbody ¾–4½ in. wide and 1½–4½ in. tall. Young branch tips dotted yellow to completely yellow, branches light red. Odor and taste mild. Edible. Under conifers in fall in Cascadia. Young *R. cyaneigranosa* var. *persicina* branches are peach- to salmon-colored, tips dotted yellow. The genetically distinct *R. cyaneigranosa* var. *elongata* (< 1½ in. wide and 5 in. tall [page 318]) with peach-colored to pastel red branches and reddish branches tips, needs elevation to species rank.

↓138b. Branches and tips scarlet when young, soon fading to orangish

Ramaria stuntzii

Fruitbody < 6 in. wide and 7 in. tall. Texture soft and stringy. Odor mild and taste slightly bitter. Usually under hemlocks but also other conifers in the fall in Cascadia. Indistinct odor, taste slightly bitter. Edible. Stipe flesh blues with Melzer's reagent.

↓138c. Branches red to crimson; fruitbody < 3 in. wide and 5 in. tall

Ramaria araiospora var. *rubella*

Ramaria araiospora var. *araiospora*

In variety *rubella*, branches and tips remain crimson to magenta. In variety *araiospora*, branches and tips fade when exposed to light. Softly stringy texture. Mild odor and taste. Edible. Under western hemlock. Fall in Cascadia.

138d. White branches with red or pink tips; flesh color same as surface

Ramaria botrytoides

Fruitbody < 3 in. wide and 4 in. tall. Profuse branching. Flesh firm, brittle. Odor and taste of green pea hulls, very bitter to peppery. Not edible due to long-lasting pepperiness. Distribution NA.

139a. (137a) Base has a "rusty root"; young branch tips deep, dull, fleshy red

Ramaria coulterae

Spring and early summer species is ≤ 4 in. tall and 5 in. wide, often growing in huge fairy rings in somewhat dry coniferous forests, often with ponderosa pine in Cascadia. Odor is indistinct, taste is mildly nutty. Edible and good.

↓139b. Massive base bruises wine-red; pink branch tips soon lose color

Ramaria rubrievanescens

Fruitbody < 4 in. tall and wide. Pinkish coloration of branch tips disappears at maturity or soon after picking, changing to white or pinkish beige. Found on the ground under conifers in spring and fall. Distinctive and choice. Eastern and western NA.

↓139c. Massive base; branch tips purple-red, slowly fading

Ramaria rubripermanens

Distinctive fungus can reach 8 in. tall and 4 in. wide. Found spring and fall above 2,000 ft. under various conifers in Cascadia. White pine is supposedly favored, but I find them under a mix of Douglas fir and grand fir, and only in the spring. The odor is weakly fragrant and the taste raw is weakly of vegetables, nutty when cooked. A choice edible.

139d. Branch tips pale red to light lilac; base may stain yellowish or tannish

Ramaria 'botrytis'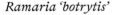   *Ramaria botrytis* PNW01

Large (≤ 7 in. tall and wide), distinctive fall species with a faintly sweet odor and mild taste. Found under conifers. A good edible. As with all edible species, a few people will suffer mild GI distress after eating. The Cascadia species is distinct enough to be at least a separate variety. Distribution NA, EU, North Africa.

140a. (2c) Toothed species with a distinct cap and stipe (page 149) *144a*

↓140b. Masses of light colored, soft spines growing from wood *142a*

↓140c. Wood-like mass of spines high on conifers; red-orange inside

Echinodontium tinctorium

Hoof-shaped fruitbodies, ≤ 10 in. wide and 6 in. tall, grow on living and dead true firs and hemlocks in western NA, including Mexico. Native Americans ground them up to use for pigment. Mixed with grease, they were used to paint faces for decoration, sunscreen, and mosquito bites. No odor or taste. Not edible.

140d. Tiny spines hanging on underside of old conifers *141a*

141a. (140d) Spines whitish with a tiny, hairy stipe

Mucronella fusiformis

M. fusiformis and the nearly indistinguishable *M. pendula* (Cascadia, Australia) have a short, hairy stipe at the point of attachment, unlike other *Mucronella* species. The spines are about ¼ in. long. Generally found on rotting conifer wood in October and November. (See also 94c [page 115].)

141b. Tiny yellow and some white spines

Mucronella 'pulchra'

The name *M. 'pulchra'* is based on material from Pakistan, and Cascadia species may not be the same. Spines are tiny, less than ¹⁄₁₀ in. long. A white species, *M. calva* (= *M. bresadolae*?), is similar and can cover vast parts of the undersides of old, rotten conifers. Distrubition NA, EU.

142a. (140b) Small (⅕–1 in. long) teeth on a branching structure arising from a small point of attachment to dead tree *143a*

142b. Large (½–3 in. long) teeth arising directly from a broadly attached, hairy mass of tissue from wounds on ± living hardwoods

Hericium erinaceus

My favorite edible is ≤ 16 in. tall and wide, found on wounds in living oak trees and on old dead hardwoods in NA, EU, and Siberia from October to December. Wild specimens taste like lobster, mild when cultivated. Best while still whitish. Potential medicinal mushroom.

143a. (142a) Teeth ± evenly distributed along branches

Hericium 'coralloides'

Not frequently encountered. Fruitbody < 14 in. diameter. Does not tend to be meaty or massive, and flavor is not great, but it is still a much better than average edible. On hardwoods in NA, EU. The Cascadia species may be genetically distinct.

Cottonwood log with *Hericium americanum* (below)

↓143b. Teeth clustered at the branch tips, usually on hardwood

Hericium americanum

Fruitbody ≤ 12 in. tall and wide. Rarely on conifers. On living as well as dead trees, late summer and fall in NA. Mild odor and taste. Decent edible with potential medicinal value. Larger spores than *H. abietis*.

143c. Teeth clustered at branch tips; always on dead conifers

Found in late summer through fall at all elevations in western NA. Commonly known as bear's head. A single dead tree can have from one to a dozen specimens. Size ranges from baseball size to bushel basket size. Rivals my favorite, *Hericium erinaceus* (lion's mane), as the best tasting of all wild mushrooms. Grew by the thousands on blowdown from Mount St. Helens eruption. Look for logs with at least one end in a wet area. Common on logs at sawmills. Of medicinal interest.

Hericium abietis (both images)

144a. (140a) With a cap and stipe; flesh too tough to eat (page 152) *150a*

144b. With a cap and stipe; flesh soft like a chanterelle *145a*

145a. (144b) Cap and stipe; colors dirty orange to brown or violaceus (page 150) *146a*

↓145b. Cap light orange-brown, < 2 in. wide; center ± umbilicate, ± slender

Hydnum 'umbilicatum'

The species in the photo is probably *H. oregonense*, a newly named species with slightly larger spores than a yet unnamed look-alike to *H. umbilicatum* (EU). *H. melitosarx* has small spores and is rarely depressed in the center. These are called sweet tooth and are in my top ten favorite edibles. Another look-alike is in eastern NA.

145c. Cap (1–11 in. wide), white to light orange brown, not slender

Hydnum washingtonianum (= H. neorepandum)

We probably have at least three species in Cascadia that have been referred to as *Hydnum repandum* (EU). The third species is *H. melleopallidum*, < 1½ in. wide. They are all edible, with a texture like their close relatives, the chanterelles. They are usually sweeter and more flavorful than any chanterelle, though they can at times be bitter when large (they can reach dinner plate size). Found late summer through fall under conifers. On tooth fungi, the "teeth" start short and grow longer and longer with age.

Hydnum cf. *olympicum*

146a. (145a) Cap ± smooth when young, cracked or a bit scaly in age *149a*

146b. Cap rough to distinctly scaly when young *147a*

147a. (146b) Cap and flesh not violet *148a*

147b. Cap (2–7 in. wide); cap and flesh violet

Hydnellum fuscoindicum
(= Sarcodon fuscoindicus)

Dark violet coloration of all parts is distinctive. Under hemlock and pine in late summer through fall in western NA. Odor and taste are mildly of cinnamon or mildly to strongly farinaceous (starchy). Not recommended.

148a. (147a) Cap ± flat, zoned, and ± cracked, whitish, brownish, and orangish

Hydnellum stereosarcinon
(= Sarcodon stereosarcinon)

Common in the spruce-fir zone in late summer and fall in NA. Can be 12 in. across. Odor and taste are mild to slightly farinaceous. Distinctive but not edible.

148b. Cap (≤ 8 in. wide) convex to umbilicate, scaly even when young

Sarcodon imbricatus group   *Hydnellum scabrosum*

The *S. imbricatus* group (three to four species) is common under spruce and firs in Cascadia. Scales are large, coarse, truncate. Odor sour. Taste mild to bitter. Edible. *S. squamosus* (under pines) is a dye mushroom. *H. scabrosum* (under spruce) and *H. glaucopus* (under pines) are distinguished by bluish green flesh in stipe base. Edible? In Cascadia, EU.

149a. (146a) Cap pale pinkish brown, purplish brown, or grayish brown

Phellodon 'violascens'

Whitish spines, a white spore print. Fruitbody < 4 in. wide. Either three sister species or one highly variable species under conifers and mixed woods in late summer to fall in NA, EU. Mild odor and taste. Edibility? Previously known as *Bankera violascens*.

↓149b. Grayish to dark brown cap (< 6 in. wide); cut flesh turns yellowish green

Hydnellum 'versipelle'

Can have a reddish margin. In age, surface breaks into small appressed scales. Under mixed conifers and hardwoods, late summer through fall in NA, EU. Odor spicy, taste ± peppery. Edibility? Previously known as *Sarcodon versipellis*. The Cascadia species may be a unique sister species.

149c. Cap flesh vinaceous-buff, unchanging; ± violet tint to stipe base flesh

'Sarcodon' rimosus

Mature cap (< 5 in. wide) becomes cracked and a bit scaly. Found in mixed forests in the fall in western NA. The odor and taste are mild to farinaceous, sometimes peppery. Edibility? Will be moved to genus *Hydnellum*.

150a. (144a) Fruitbody top-shaped, thick flesh; brown spores *152a*

↓150b. Fruitbody top-shaped, thin flesh; white spore print *151a*

150c. Small, fuzzy cap; long, slender, fuzzy stipe; on a conifer cone

Auriscalpium 'vulgare' CA01

This distinctive little species (about 1 in. wide and ≤ 4 in. tall) is easy to overlook because of its small size and dark brown cap. Most often on Douglas fir cones, sometimes on ponderosa pine and spruce cones. On old, buried cones, it appears terrestrial. Of interest for its uniqueness and beauty. Not an edible. *A. vulgare* (EU) has a central stipe and 3 percent different DNA. The western NA species will need a new name.

151a. (150b) Cap, flesh, and stipe bluish black to purplish black; gray teeth

Phellodon atratus

Caps are ½–2 in. across. Adjacent caps often fuse. Under spruce, true fir, Douglas fir, and hemlock CA, OR, BC. Visible year-round, but actively growing late summer through fall. Odor mild to slightly fragrant, taste mild. Inedible.

↓151b. Caps grayish brown to ± purple-brown, with ± yellow-brown tones

Phellodon melaleucus

Caps (< 3 in. wide) fuse in groups, < 8 in. wide. Flesh brown to purplish gray or purplish black, stains olivaceous in KOH. Under conifers and in mixed woods in NA, EU. Aromatic odor, mild to bitter taste. Seen all year, active late summer to fall.

151c. Fuzzy cap white when young, soon yellow-brown to brown

Phellodon tomentosus

Caps (< 2 in. wide) fuse. Flesh is two-toned brown, blackening in KOH. Whitish teeth bruise vinaceous buff. Actively growing late summer and fall under conifers in NA, EU. Odor aromatic, taste mild to ± bitter. Inedible.

152a. (150a) Flesh in lower stipe interior orange to buff (page 154) *153a*

152b. Cap (< 12 in. wide) starts out white; flesh in lower stipe bluish

Hydnellum 'suaveolens'

Develops yellowish to tan zones, finally violet-gray from the center outward. Odor ± fragrant (anise-like), the taste somewhat bitter. Late summer through October, under conifers, especially hemlock in NA, EU. The Cascadia species is at least closely related. A dye mushroom.

153a. (152a) Cap fuzzy, violaceous black, concentric ridges; forms rosettes

Hydnellum regium (very young)   *Hydnellum regium* (mature)

Caps (≤ 4 in. wide) fuse in groups, to 8 in. wide. Flesh violaceous near the cap and dull tan at the base of stipe. The lower stipe exterior is a pinkish cinnamon color. New growth first appears in August, and it fruits through October under true firs, hemlock, and alder in Cascadia. Odor is heavily aromatic and ± farinaceous. Taste is disagreeable. Inedible. Highly prized for fabric dyes.

↓153b. White to pink cap surface, ± blood-red droplets

Hydnellum peckii (above and at right)

Fruitbody ≤ 6 in. wide. Actively growing August to October under conifers in NA. Cap initially white to pink, darkens from the center outward to brown-vinaceous or brown with age. Blood-red droplets disappear in dry weather. Odor is fragrant to pungent, taste very peppery. All *Hydnellum* species are of interest to fabric artists, who value them for fabric dyes. *H. peckii* is common and is prized. *H. regium* is not common except in the Columbia Gorge (OR, WA) area, where it is gathered in quantity by fabric artists.

↓153c. Cap pale blue or whitish, aging dark brown from center outward

Hydnellum caeruleum

Cap ≤ 7 in. wide, orange stipe base. Found under conifers and sometimes hardwoods starting in August, actively growing through October in NA, EU, Asia. Odor and taste ± farinaceous.

153d. Cap (< 6 in. wide) orange to greenish cinnamon color; orange to carrot-red flesh

Hydnellum aurantiacum (above and at right)

The image at right was formerly known as *H. complectipes*, which is now a synonym of *H. aurantiacum*. Abundant August to October under conifers in NA, EU, Asia. In addition to *H. aurantiacum*, two unnamed sister species are present in Cascadia. The odor and taste are ± farinaceous. Inedible. *H. auratile* (western NA, EU) is distinguished by smaller spores.

154a. (2b) Greenish yellow to orange-colored jelly fungi (page 158) *160a*

↓154b. Brown to black jelly fungi *157a*

↓154c. Clear to whitish, or slightly ochraceous jelly fungi *156a*

154d. Pink gelatinous fungi *155*

155. (154d) Pink gelatinous ascomycete in clusters ≤ 4 in. across

Ascocoryne cylichnium/A. sarcoides

These two beautiful, little (< ½ in. diameter), pink, jelly-like mushroom species are differentiated only by spore size. They grow on stumps and logs globally. Odorless, tasteless. (See also key lead 2b #1 [page 42].)

156a. (154c) Distinct stipe; cap with tiny, soft teeth on the underside

Pseudohydnum 'gelatinosum' group (two species, ± life-size)

This beautiful little fungus comes in a clear-topped version and a brownish-topped version. They have a rubbery or gummy bear-like consistency. Widespread in coniferous woods in summer and fall. Odorless and tasteless. Harmless. Sometimes eaten with honey and cream. Global distribution. One Cascadia species is close to the EU species.

156b. Fruitbodies whitish blobs or brain-like, reddish brown in age

Exidia candida

Fruitbodies ≥ 4 in. wide are broadly attached. Primarily observed late in the season and during mild, wet winters. Grows on a range of hardwoods in NA. Like all jelly fungi, virtually all water and no flavor, so of no interest as edible.

157a. (154b) Small to medium-small fruitbody; on wood *159a*

157b. Medium-large to large fruitbody; on the ground *158a*

158a. (157b) Large (≤ 8 in. wide); very gelatinous interior

Urnula padeniana

Formerly known as *Sarcosoma mexicanum* (a misapplied name), this distinctive spring ascomycete grows under conifers at all elevations from BC to Mexico and in India. It is odorless and tasteless.

158b. Medium-large; barely gelatinous interior

Pseudosarcosoma latahense

Fruitbody ≤ 3 in.wide, very gelatinous only when young. Formerly known as *Sarcosoma latahense*, this spring ascomycete fruits April to June on the ground or on rotting wood at all elevations in Cascadia, often just after snowmelt. It is rubbery, odorless, and tasteless.

159a. (157a) Fruitbody gray-brown, < 1 in. wide; ± fusing masses ≤ 20 in.

Exidia nigricans/E. glandulosa group

E. nigricans, considered by some as a synonym of *E. glandulosa*, produces individual fruitbodies that quickly coalesce, while *E. glandulosa* fruitbodies remain separate. Common on hardwoods, especially oaks. Rare on conifers. Global distribution. Not an edible.

159b. Fruitbody (< 6 in. wide) brown and ± ear-shaped

Auricularia americana and *Auricularia* cf. *angiospermarum*

A. americana (see key lead 2b #2 [page 42]) is on conifers and *A.* cf. *angiospermarum* (from MD) is on hardwoods. The Cascadia species is unnamed. Edible. Late summer to winter. Global distribution (as *A. auricula*).

160a. (154a) Fruitbodies yellow-orange to bright orange *162a*

160b. Fruitbodies yellow and/or with greenish tones *161a*

161a. (160b) Stipe present; round, gelatinous; yellow to pale orange or green head

Leotia 'lubrica'

Viscid, gelatinous species is ± ½ in. wide by 2 in. tall. Found occasionally, ± year-round in hardwood and coniferous forests in NA, EU. The Cascadia species is unique and unnamed. The color is yellow, greenish, or olive-green. The green-capped variety was once known as *L. viscosa*. No odor or taste. Not edible.

↓161b. Small clumps (< 1 in. wide), yellow-green globular mass; on hardwood

'Dacrymyces' tortus/ 'Dacrymyces' minor group

'D.' minor and *'D.' tortus* are distinguished by microscopy. Found year-round in wet weather as a global species complex. *'D.' tortus* is mostly on conifers and *'D.' minor* is mostly on hardwoods, both on oaks. *Dacrymyces stillatus* (global) and *D. minutus* (NA) are not greenish and are separated microscopically. Expect changes in genera and species globally. (See key lead 166b [page 160].) Odorless and tasteless.

161c. Yellow to yellow-orange gelatinous fungus; on hardwoods

Tremella mesenterica

Fruitbody ½–4 in. wide and ± 1 in. tall. Parasitic on *Peniophora* species (crust fungi) that grow on hardwoods. Spring to fall in wet conditions, globally. Of no interest as an edible. Color varies: clear, yellow, or pale orangish. *T. mesenterella* (NA, page 318) is distinguished microscopically.

162a. (160a) Growing on wood or a wood-loving fungus (see also key lead 161c) *164a*

162b. Growing on ground or from buried wood; stipe present *163a*

163a. (162b) Resembles a small, rubbery coral mushroom, ≤ 4 in. tall

Calocera 'viscosa' (both images)

Firm and gelatinous, arising from a stout, white rooting base in the soil or on rotten coniferous wood. Fruits August through November in Northern Hemisphere. The one or two Cascadia species are unique. Tasteless and odorless. Not edible. Branches of coral mushrooms break when bent.

163b. Resembles a 1–5 in. tall, colored funnel split down one side

Guepinia helvelloides (three images)

Formerly *Phlogiotis helvelloides*. Distinctive. Odorless, tasteless, harmless. Under conifers August to December. Distributed globally. May be found in Haida Gwai, an archipelago on Canada's Pacific coast. Elsewhere in Cascadia the species is unique and unnamed.

164a. (162a) Growing directly on wood; not a parasite (page 160) *165a*

164b. Parasitizing a *Stereum* species; several clear, yellow, orange, or brown species

Naematelia aurantia (= *Tremella aurantia*)

N. aurantia is translucent to pale orange with a white core, found mainly on hardwoods, as is *Phaeotremella frondosa* (with brownish leaf-like folds). *N. encephala* (brain-like, white to pinkish brown or yellowish) is found on *Stereum sanguinolentum* on conifers, as is *P. foliacea*, a brownish leaf-like species. All are odorless and tasteless, found in winter and spring in many parts of the world.

165a. (164a) Small top-shaped, pointed, or spathulate fruitbody *167a*

165b. Small rounded, globular, or brain-like fruitbody *166a*

166a. (165b) Small yellow cushions attached by a tiny stipe

Dacrymyces capitatus (± full size)

Grows year-round on hardwoods, rarely on conifers, worldwide. Shaped like a tiny top, ± ¼ in. wide. The small rooting stipe distinguishes it from the jellies that grow as more broadly attached blobs on wood. Too tiny to eat. *D. stillatus* lacks a stipe.

↓166b. Light yellow to orange easily squashed cushions that deliquesce (liquify) in age

Dacrymyces stillatus (± full size)

On conifer branches, rarely hardwoods, worldwide. Blobs can coalesce into a large mass. Found year-round during wet periods. Of no interest as an edible. Similar species include *'D.' tortus*, *'D.' minor*, and *D. capitatus*. If a white rooting base is present, see *D. 'chrysospermus'*.

166c. Bright orange, ± brain-like; tough; white rooting base

Dacrymyces 'chrysospermus'

Fruitbody in masses, > 2 in. diameter. Formerly *D. palmatus*. Found on both hardwoods and conifers, worldwide. Two unnamed sister species are known from Cascadia. The tough, white rooting base distinguishes it from similar jelly fungi. Harmless if eaten. *D.* cf. *aquaticus* (see key lead 2b #3 [page 42]) is on very wet, usually coniferous wood.

167a. (165a) Fruitbody < ¾ in. tall; clustered yellow to orange clubs

Calocera cornea

Common on barkless hardwood, sometimes conifers, summer and fall worldwide. If tip is palmate to spathulate and one side is paler than the other, it is a different global species, *Dacryopinax spathularia* (see key lead 2b #4 [page 42]). Both are tiny, odorless, and tasteless. Both resemble a small *Calocera viscosa* (see key lead 163a [page 159]).

↓167b. Fruitbody top-shaped; on conifers not long after snowmelt

'Heterotextus' 'alpinus'

Widespread globally. The Cascadia species belongs in a new genus; it is a sister to *'H. alpinus'* and needs a new epithet as well. One of several ⅛–⅜ in. wide, top-shaped jelly mushrooms. *Dacrymyces chrysocomus* (dries red-brown) and *D. variisporus* (orange to orange-brown) are best distinguished microscopically, but fruit later, as does *H. luteus*. All are odorless and tasteless.

167c. Fruitbody top-shaped; grows later in the season

Heterotextus luteus

H. luteus (< ½ in. wide and ½ in. long) is not a snowbank fungus. It appears in summer in Cascadia and New Zealand. It has a tiny stipe, while the look-alikes (including *Dacrymyces chrysocomus* and *D. variisporus* [NA, EU, Asia]) are ± broadly attached. At a minimum, *D. chrysocomus* and *D. variisporus* need a new genus.

168a. (2a) Fruitbody fleshy (all are basidiomycetes) *170a*

Note: For teeth hanging on the underside of a log, see key leads 141a and 141b (page 146). For gelatinous clubs, see key lead 167a (page 161). For ascomycetes, see key leads 61a–64b (pages 88–90).

168b. Fruitbody tough, fibrous, almost woody *Xylaria* species *169a*

169a. (168b) Multiple whitish fruitbodies; on wood

Xylaria cornu-damae

Tough, wiry fruitbodies have a black exterior that looks whitewashed, only showing black where surface layer is rubbed off. Can be ≤ 3⁄8 in. tall. Growing from buried hardwood and thus appearing terrestrial in the image. Odorless, tasteless, and inedible. A rare NA ascomycete needing DNA analysis.

169b. Strap-like to antler-like, white-tipped, black fruitbodies; on wood

Xylaria hypoxylon

Black, hairy, < 3 in. tall stipe grows from black, hairy pad, typically on hardwoods, sometimes conifers, in Cascadia. White tip is due to asexual spores. Antler-like branching is common (see last image in key lead 2a [page 42]). Odorless, tasteless, and inedible. A very common year-round ascomycete.

170a. (168a) Clustered purple clubs

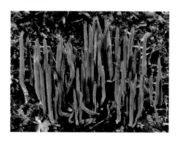

Alloclavaria purpurea group

Very common, fragile, purple club (5 in. tall) can literally carpet the ground under conifers, growing in cespitose tufts in NA and EU. In Cascadia, at least one look-alike is also present. Odor and taste vary from mild to nauseating, though it is edible. Also known as *Clavaria purpurea*. *Clavaria rosea* (NA, EU) is pink, rare, and ≤ 2 in. tall. (See *C. rosea* illustration in key lead 2a #1 [page 42].)

↓170b. Small (≤ 1 in. tall), trumpet-shaped clubs

Clavicorona taxophila

Starts white and turns yellowish in age. On conifer debris and hardwood leaves, typically near yew trees. No reports on odor, taste, or edibility (too small anyway). Found during fall and early winter season in NA, EU.

↓170c. Not slender and white or small and ± white overall (page 164) *172a*

170d. Clubs ± white, either small diameter or short and small *171a*

171a. (170d) A clustered mass of ≤ 6 in. tall; slender, white clubs

Clavaria fragilis group

Long known as *C. vermicularis*. Starts out as a cluster of short clubs (see key lead 2a #2 [page 42]). Distinctive. Found in summer and fall under conifers, especially spruce. Often grows in large masses. Odor mild, bland taste. Edible. Global distribution.

171b. Tip truncate with a central pointed projection; ≤ 3 in. tall

Clavariadelphus mucronatus

Whitish when young, aging tan to pinkish. On conifer debris in the fall and early winter in Cascadia. No reports on odor, taste, or edibility (too small anyway). The ± truncate end with a central point is distinctive. (See also 174b [page 165] for similar species lacking the mucro [central pointed projection].) The entire *Clavariadelphus* genus globally needs study and revision.

172a. (170c) Small and/or slender; single to gregarious *174a*

↓172b. Medium-sized; single to gregarious but not cespitose

Clavariadelphus truncatus

Fruitbody 2–7 in. tall and 1–4 in. wide. Can be yellowish orange, pinkish brown, or orangish brown. (See also key lead 2a #3 [page 42].) Top is club-shaped to truncate. Found under mixed conifers. Common. Odor none, taste sweet to bitterish. Edible and prized by some. Globally distributed.

172c. Medium-sized; growing in cespitose clusters of two or three *173a*

173a. (172c) Clusters of two or three large, white to pallid clubs, 2–10 in. tall and ≤ 1½ in. wide

Clavariadelphus caespitosus

Off-white to light yellow fruitbodies bruise brown. Found in mixed woods in fall and winter, rarely spring. No distinct odor or taste. Edibility? One of three species in Cascadia previously known as *C. pistillaris*.

173b. Cluster of 2–10 in. tall, grayish pink clubs; bruising brown

Clavariadelphus occidentalis

Once lumped under *C. pistillaris*, along with the even darker *C. subfastigiatus*. *C. caespitosus* ages to pinkish cinnamon or vinaceous brown. Found in both coniferous and mixed woods during the fall and winter. No distinctive odor or taste and no reports on edibility. The flesh of *C. subfastigiatus* turns green with a drop of KOH solution, while *C. occidentalis* and *C. caespitosus* are unchanged. Found on NA West Coast from AK to Mexico. The entire *Clavariadelphus* genus globally needs study and revision.

174a. (172a) Single to small tufts of slender yellow to orange clubs

Clavulinopsis laeticolor

Found year-round in mild weather on leaf mold or humus under hardwoods or conifers in NA, EU, Asia. Distinctive lemon-yellow to bright orange, ½–4 in. tall. No notable odor or taste. Bright yellow *C. fusiformis* tends to fruit in small, fused clumps and is bitter tasting. KOH turns *C. laeticolor* flesh greenish but does not affect *C. fusiformis*.

↓174b. Dull yellowish to dull orange brown; bruising darker

Clavariadelphus ligula

Under mixed conifers in late spring through fall in NA, EU, Asia, Africa. Can be ≤ 4 in. tall and ½ in. wide. Odor none, taste mild, ± sweet, or ± bitter. Not worth eating. *C sachalinensis* differs microscopically.

↓174c. Gregarious dull orange clubs (< 1 in. tall) associated with algae

Multiclavula vernalis

Basidiolichen. Creamy to dull orange-brown club, translucent stem. Growing in moss with green algae in the spring (NA, EU). *M. corynoides* (NA, EU) is a look-alike. *M. mucida* is similar but is on algae-covered old wood, found globally. No odor, moldy taste. Inconsequential.

174d. Tough, threadlike club; on wood

Macrotyphula fistulosa

Can be ≤ 8 in. tall but so threadlike that the ochre-yellow to brownish clubs are often unnoticed. Grows on both hardwoods and conifers, especially alder, in fall and winter. No odor, no taste, no substance. Some authors place it in the genus *Typhula*. Found in cooler parts of Northern Hemisphere.

Note: The remainder of this key deals with the gilled mushrooms. To identify them, it is critical that you know the spore color by taking a spore print or observing spores deposited in the field on underlying mushroom caps, underlying vegetation, or on the mushroom stipe. With practice, you will come to recognize most gilled genera on sight, without needing to take a spore print. It helps to cut the mushroom in half from top to bottom to see the gill attachment clearly.

175a. (1a) Spores dark chocolate-brown to black (page 293) *385a*

↓175b. Spores light brown, orange-brown, or rusty brown (page 257) *327a*

↓175c. Spores pinkish salmon (page 251) *315a*

↓175d. Spores white or olive-green; gills free from stipe *177a*

175e. Spores white, cream, pale yellow, pale orange, or pinkish buff; gills attached to stipe (with a notch, bluntly or decurrent) *176a*

176a. (175e) Growing on wood (or from buried wood) (page 245) *305a*

↓176b. Breaks crisply (*Russula* and *Lactarius* spp.) (page 229) *276a*

↓176c. Gills thick, often brightly colored; sometimes viscid cap; fleshy stipe (*Hygrophorus* and *Hygrocybe* spp. and relatives) (page 219) *262a*

↓176d. Caps usually < 1 in. wide, often conical; fragile cap and stipe; no partial veil (Mycenoid) (page 213) *253a*

↓176e. Caps < 1 in. wide, may be depressed to umbilicate; sometimes decurrent gills; no veil; thin stipe (Omphalinoid) (page 207) *240a*

↓176f. Caps small to large, convex to flat; gills generally bluntly attached and usually closely spaced (Collybioid) (page 200) *223a*

↓176g. Caps > 1 in. wide; may have a partial veil, sometimes viscid; gills usually notched; fleshy stipe, (Tricholomatoid) (page 190) *207a*

176h. Caps > 1 in. wide, like Omphalinoid; adnate to decurrent gills, rarely viscid; rarely veiled (Clitocyboid) (page 180) *196a*

177a. (175d) Gills distinctly free; cap breaks from stipe in a ball-and-socket fashion; sometimes scaly, not viscid (*Lepiota* spp. and allies) (page 174) *190a*

↓177b. Gills barely free of stipe; cap does not separate from stipe in ball-and-socket fashion; membranous universal veil present (*Amanita* spp.) *178a*

177c. Gills barely free of stipe; viscid universal veil (page 179) *195a*

178a. (177b) Volva not a membranous cup (page 169) *185a*

178b. Volva a membranous cup *179a*

179a. (178a) Partial veil thin or absent; prominent cap striations (page 168) *182a*

179b. Thick and membranous partial veil (and universal veil) *180a*

180a. (179b) Fruits late summer to winter under oaks and other hardwoods

Amanita phalloides

Accidental introduction from EU, found on the NA East Coast and West Coast, from Los Angeles to Vancouver, BC, as well as many other parts of the world. Cap can be white, greenish, or bronze; ± volval patch on cap. Delicious but DEADLY!

180b. Fruits late winter to spring in mixed woods, usually with oaks *181a*

181a. (180b) Volval patch on cap; cap 2–6 in. wide, ± white when young; sometimes striate

Amanita ocreata group

At least two species from Southern CA to WA. Prefers low elevations, often sandy soils, mixed woods, often with cottonwoods or oaks. Mild odor when young, aging ± fishy. Do not taste. Delicious but DEADLY! Has been mistaken for *A. calyptroderma*, *A. vernicoccora*, and *A. velosa*.

181b. Volval patch on cap; cap 2–7 in. wide, ± pale yellow when young; faintly striate

Amanita vernicoccora

Cap pale yellow-orange to almost white, typically with a large patch (remains of the universal veil) and faint striations on the cap margin. The volva is a broad cup. *A. calyptroderma* (page 36) is more orange to brownish (but has a white form) and fruits from fall to winter. Both odor and taste mild, aging ± fishy. Both species are edible but NOT recommended (two deadly look-alikes). Range CA to BC.

182a. (179a) Partial veil absent *183a*

182b. Thin partial veil leaving a unique ± dissolving ring on stipe

Amanita calyptratoides

Has a unique, watery-looking stipe from apparent dissolution of the ring. Saccate membranous volva. Viscid when wet. Found under Oregon white oak in Klickitat County, WA, known from live oaks in Southern CA. Odor of wet soil, taste mild. Edibility?

183a. (182a) Volva constricted around the lower stipe, flaring above

Amanita constricta

Cap (≤ 5 in. wide) is gray with prominent striations, sometimes with buff to gray universal veil remnants. The constricted cup can be tan to gray inside, sometimes reddish stained outside. Uncommon, fruits in the fall under oaks and Douglas fir, CA to BC. Odor and taste not distinctive. Edible, but not choice. Be certain of your identification. Several other *Amanita* species may have a constricted volva.

↓183b. Fruitbody ± white; volva disintegrates leaving fragments on the cap, lower stipe; cap margin striate

Amanita obconicobasis n.p.

Found in CA and under oaks in Klickitat County, WA. Indistinct odor. Toxic? Both the deadly *A. ocreata* and *A. velosa* (pale coral-pink to peachy pink cap coloration) are similar.

183c. Fruitbody not white, volva not constricted, partial veil absent *184a*

184a. (183c) Brown cap, brown band near striations

Amanita pachycolea group

A group of similar species found from CA to BC. The ≤ 6 in. wide cap varies from gray-brown to dark brown; gills fringed with dark brown; volva thick and rust stained. Occasionally under conifers and in mixed woods. Odor mild to unpleasant, taste mild to fishy. Edible but not choice. An ID mistake could be deadly.

184b. Cap gray to tan, no brown band

Amanita 'vaginata' group

Amanita 'fulva'   *Amanita friabilis*   *Amanita lindgreniana* n.p.

A. 'vaginata' group members share distinct striations on the cap margin, a non-constricted volva, tall and slender stature, absence of a partial veil, no ring on the stipe. Found in NA, EU, Asia, North Africa. A. lindgreniana n.p. and A. friabilis grow in sandy soil under mixed conifers near mountain alder. A. 'fulva' grows at low elevations under conifers and hardwoods. All are edible, but not choice. Similar toxic species.

185a. (178a) Partial veil present when young; ± ring on stipe (page 170) *186a*

185b. Partial veil absent, powdery universal veil; cap 1½–3 in. wide

Amanita 'farinosa'

No ring or ring zone on the stipe. Base slightly bulbous. Universal veil leaves powder on the cap and the stipe base. In coniferous woods, August through fall. Odor indistinct. Taste and edibility not recorded. The western NA species is distinct and unnamed. The eastern NA species is one-third as large.

186a. (185a) Universal veil membranous; rolled collar volva (page 172) *189a*

↓186b. Universal and partial veils membranous; volva roughly three bands of tissue around top of a bulbous base *188a*

186c. Universal and partial veils powdery, disappearing in age *187a*

187a. (186c) Powdery white cap and short white stipe, ± marginate stipe base

Amanita silvicola

The 2–5 in. wide cap surface is cottony to powdery, the gills are white with powdery edges, and the stout stipe is powdery or with cottony scales. Common from August through fall under conifers in Cascadia. Odor mild, fishy in age; taste is mild. Edibility? Dangerously similar looking to poisonous *A. smithiana*.

187b. Powdery white cap and long white stipe, stipe base swollen

Amanita smithiana (above and at right)

Cap 2–7 in. wide; gills white with powdery edges. Stipe has a large spindle-shaped base and powdery partial veil. Under conifers in Cascadia. Odor ± bleach-like; taste mild. Toxic. Causes kidney failure if eaten, but kidney function slowly recovers (over months) and no known deaths to date. Often mistaken for prized and edible *Tricholoma murrillianum* (white matsutake, key lead 211b [page 192]).

188a. (186b) Universal and partial veils yellow to grayish

Amanita augusta

Cap 1½–5 in. wide, dark brown to yellow-brown; gills white, often with a dark edge. Stipe 1½–5 in. long with enlarged base. Late summer through fall under conifers in Cascadia. Odor and taste mild. Probably toxic. Formerly known as *A. aspera* and *A. franchetii*.

↓188b. Universal veils and partial veils white; cap 1½–4 in. wide

Amanita aprica

Yellow to orange cap has tightly adhering universal veil remnants; membranous annulus is low on the stipe. Common in coniferous woods at all elevations in the spring and early summer in Cascadia. Odor and taste not distinctive. Poisonous.

188c. Veils creamy white; fall to winter

Cap can be ≤ 10 in. across and usually red to orange (see key lead 1a [page 42]), but can be yellow or white. Volva forms a distinctive set of rings at the top of the bulbous base. Membranous

Amanita muscaria subsp. *flavivolvata*/*A. chrysoblema*

annulus. *A. chrysoblema* and the look-alike *A. muscaria* subsp. *flavivolvata* are found throughout NA under conifers and hardwoods. *A. muscaria* var. *muscaria* (a pure white veil) is found in EU and AK and sometimes grows near NA cities. Toxic, mind-altering (as are all species in the group).

189a. (186a) Grayish brown cap; bulbous base typically with a deep cleft

Amanita porphyria (above and at right)

Cap 2–6 in. wide, dark gray to dark brown with grayish warts or patches; stipe with a marginate bulb. Under conifers and mixed forests in fall, common. Odor of raw potatoes, taste not reported, probably toxic. Distribution NA, EU, Asia, North Africa.

↓189b. Light to dark brown cap; bulbous base with rolled-collar volva

Amanita pantherinoides (above and at right)

Formerly known as *A. pantherina* (EU), this species can be found from spring to fall under both conifers and hardwoods at all elevations in Cascadia. Color is variable, and yellowish forms can be a challenge to distinguish from the yellow *A. gemmata*, with which it may hybridize. The odor and taste are indistinct. When eaten, causes generally unpleasant mind-altering effects; a violent, yet comatose state, and severe gastrointestinal distress. The darker the cap, the higher the toxin levels.

189c. Yellow to yellow brown cap, marginate volva *Amanita gemmata* group

Amanita gemmata

Amanita 01mwb110613

Amanita cf. *pseudobreckonii* n.p.

A. gemmata fruits late summer to fall in many types of woods in NA and EU. The cap can be 1½ in. to 5½ in. wide. Indistinct odor and taste. Same toxins as *A. muscaria* and *A. pantherinoides*. It can be difficult to distinguish from yellowish shades of *A. pantherina* and from the genetically different but very similar toxic species, *A. pseudobreckonii* n.p. and *A. breckonii*, which fruit in the same habitats and seasons.

190a. (177a) White spore print *191a*

190b. Olive-green spore print; cap ≤ 12 in. wide

Once restricted to warm sites of NA (plus South America and the Pacific), it is expanding northward and has made it to WA and southeastern Canada. Initially smooth, surface breaks into brownish scales. Often mistaken for an edible *Chlorophyllum* species. GI distress can be severe if eaten, including bloody vomit and stools.

Chlorophyllum molybdites

191a. (190a) Cap diameter ≤ 3 in. (page 176) *192a*

↓191b. Cap diameter ≤ 12 in., smooth or scaly

Chlorophyllum brunneum

Chlorophyllum olivieri *Chlorophyllum rhacodes*

C. brunneum (marginate bulb, CA to BC) is the most common of the three choice edible species. The other two are *C. olivieri* (Cascadia), with olive-drab scales on a non-contrasting cap, and *C. rhacodes* (NA, EU), which has a bulbous, non-marginate stipe base. All grow in woods, in gardens, and on compost piles. All three bruise orange and then slowly brown. Spore print to avoid the highly similar and toxic *C. molybdites*, which bruises brown directly.

191b. (continued) Cap ≤ 12 in. across, smooth or scaly

Lepiota aspera group
(above and middle)

Leucoagaricus americanus

Leucoagaricus barssii

Leucoagaricus leucothites (now
Leucocoprinus leucothites?)

In Cascadia, the *L. aspera* (= *Echinoderma asperum*, = *Lepiota 'acutesquamosa'*) group has ± four species < 4 in. wide, with pointed scales and usually a fleeting ring (NA, EU). Indistinct to unpleasant odor and taste. Edible? *L. americanus* (< 6 in. wide, NA) is distinctive. It bruises yellow to orange and then turns reddish; the stipe is usually spindle-shaped. Often found on old sawdust piles. Odor and taste indistinct. Edible. *L. barssii* (< 6 in. wide) is a distinctive NA, EU species. The gills are close (see photo for *Chlorophyllum brunneum* [opposite]). The initially gray cap with radial fibrils can turn scaly and wood-brown in age. Found on cultivated ground, on compost, in grass. Pleasant odor and taste. Choice edible. Late summer to fall. *L. leucothites* (formerly *Lepiota naucina*, NA) has a smooth white to gray cap (diameter 2–6 in.), smooth white stipe, and a membranous ring. *L. leucothites* is easily confused with deadly, all-white *Amanita* spp., but it lacks a universal veil. Grows from late summer to fall in grass and disturbed areas. Odor and taste mild. Edible, but many suffer GI distress after eating. At least one death after someone misidentified it and actually consumed a member of the deadly *Amanita bisporigera* group.

192a. (191a) Cap scaly at least in age *194a*

192b. Cap powdery, granular, mealy *193a*

193a. (192b) Typically found in greenhouses or in indoor flowerpots; bright yellow

Leucocoprinus birnbaumii

Distinctive, small, thin-fleshed, all-yellow mushroom can be found in any season in indoor habitats. The odor is mushroomy, taste is mild. Technically edible, but not recommended. Several other *Leucocoprinus* species that favor greenhouses are distinguished by color.

193b. All white, with mealy whitish to brownish scales on the 1–3 in. wide cap

Leucocoprinus cepistipes PNW01

Like other *Leucocoprinus* species, cap is striate and the ring on the stipe is fragile and easily disappears. Grows outdoors in rich soil, compost heaps, and gardens from spring through fall. Odor is mild to fungal, taste mild to bitter. ± Edible. Distinct (DNA) from *L. cepistipes* (EU).

194a. (192a) Cap ½–1½ in. wide, with flat to pointed, reddish scales

Lepiota subincarnata

The reddish brown scales are appressed to erect, giving a ± scaly look. The flesh turns slightly red when cut. The ring is fibrillose and disappears. Found in gardens, rich soil, and woods. Odor slightly sweet-fragrant. DEADLY poisonous. Uncommon. Distribution NA, EU.

↓194b. Cap 1–3 in. wide, with reddish brown to yellowish brown scales; scaly stipe, not bruising

Lepiota clypeolaria   *Lepiota magnispora*

L. magnispora, frequent in mixed forests, tends to be smaller (1–2 in. wide) and darker than *L. clypeolaria* (1½–3 in. wide). Both have a fungoid odor and taste. *L. clypeolaria* is known to be poisonous, and *L. magnispora* looks very similar to poisonous *L. clypeolaria*.

↓194c. Cap ½–2 in. wide, with reddish brown to yellowish brown scales, sometimes with a blackish center; scaly stipe, bruising red

Lepiota flammeotincta

Cap and stipe, but not gills, bruise orange-red then dark brown. Ring disappears. Both *Leucoagaricus erythrophaeus* (smooth stipe) and *Lepiota castanescens* (fibrillose stipe) are very similar in size and color, but both bruise red in all parts. (See also *Leucoagaricus fuliginescens* PNW05 [page 318].) All in coniferous woods from fall to winter, Pacific Coast. Odor and taste unpleasant. Edibility? Distribution NA.

↓194d. Cap ½–1½ in. wide, with grayish black scales

Lepiota atrodisca group

Grows in all forest types. Five to nine look-alikes, including both *Lepiota* and *Leucoagaricus* species, are found fall to winter on soil or rotten wood in Cascadia. Indistinct odor and taste. Edibility? Very similar *Leucocoprinus brebissonii* is found spring to summer in greenhouses and on humus. Spermatic odor. Edibility?

↓194e. Cap with ± scales, pinkish brown to reddish brown center; smooth stipe

Lepiota cf. *castaneidisca* group   *Leucoagaricus 'rubrotinctus'* group

'Lepiota' cristata group

These three fall species groups are distinguished from *Lepiota clypeolaria* and *L. magnispora* in having a smooth versus a fibrillose stipe. *L.* cf. *castaneidisca* (< 1½ in. wide) group members have a sheathing, movable ring with a dark underside, and a strong, sweet-fruity odor (CA, Cascadia). *Leucoagaricus 'rubrotinctus'* group members are < 5 in. wide with an indistinct odor and taste (NA). *'Lepiota' cristata* group members, which belong in genus *Leucoagaricus*, are ¾–3 in. wide, orange-brown rather than reddish brown. Odor and taste disagreeable. Poisonous? The *L. cristata* group found in NA, EU, China.

194f. Cap ½–2 in. wide; smooth, yellowish tan center and white margin

Lepiota sequoiarum

The sheathing, flaring ring is moveable and collapses in age. Stipe is bald. Odor and taste are unremarkable. Fall species found in woods and planted areas of Cascadia. Sampled once by someone high on halucinogenic mushrooms. He survived but needed hospitalization.

195a. (177c) White- to brown-tinged, viscid cap and stipe; ± fibrillose ring zone

Zhuliangomyces 'illinitus' (= Limacella 'illinita')

Cap 1–3½ in. wide, broadly umbonate in age, margin often with hanging, viscid veil remnants. Stipe viscid below partial veil zone; partial veil fibrillose but covered with layer of slime. Odor indistinct to farinaceous, taste indistinct. Edibility? Summer to fall in wide habitat range, NA, EU, North Africa. Cascadia species is genetically distinct from (EU) *Limacella illinita*.

↓195b. Cap viscid, reddish brown; stipe fragile, dry; veil fibrillose, disappears

Limacella glioderma group

Two unnamed species in Cascadia have caps 1–3½ in. long, broadly umbonate in age, viscid to dry, often fibrillose, and revealing pinkish flesh. Odor and taste very farinaceous. Edibility? Uncommon in woods, summer to fall. Distribution NA, EU, Asia.

195c. Cap pale under reddish brown slime; margin hanging with slime

Zhuliangomyces species (= Limacella 'glischra')

Yellow-brown to red-brown cap, 1–2 in. wide, with a low umbo when old. Gills free, closely spaced. Stipe ≤ 3 in. long, covered in gluten below the glutinous, hairy ring. Cap easily separates from stipe. Indistinct odor and taste. Uncommon. Summer and fall under hardwoods and conifers. Edible? Cascadia species has been mistaken for the eastern NA species *Limacella glischra* but is genetically distant.

Note: If your mushroom appears to have pale spores and free gills but does not have a volva (is not an *Amanita* species or a rare *Limacella* species that has a viscid volva), or it does not have a cap that breaks from the stipe in a ball-and-socket fashion (*Lepiota* species and allies), it is a species in which the gills were initially attached to the stipe (typically with notched gills that tore free as the mushroom matured).

196a. (176h) Spore print white to cream-colored or pale yellow (page 182) *198a*

196b. Spore print pale pinkish buff (*Lepista* and *Collybia* species) *197a*

197a. (196b) Small species (cap ½–2 in. wide), brownish with lilac tints

Collybia (= Lepista) tarda
(three images)

Cap is hygrophanous, turning tannish when dried. In cultivated or fertilized areas, rarely in woods. Often with a thick strand of mycelium attached at base. Found summer and fall in NA, EU, Asia, North Africa, Australia. Odor and taste are mild. Edible when well-cooked.

↓197b. Stout, with a 2–7 in. wide, lilac-tan to bluish purple cap

Distinctive species found under hardwoods and conifers and on rich ground, typically in late fall. Odor and taste pleasant. Edible, but can cause bloody vomit and bloody diarrhea if only lightly cooked and eaten. At lower elevations sites in NA, EU, Asia, North Africa, Australia.

Collybia (= Lepista) nuda
(three images)

↓197c. Cap 2–6 in. wide, pale watery buff with faint purple

Collybia (= *Lepista*) *glaucocana*

Resembles a faded *C. nuda*. Found late summer through fall under both hardwoods and conifers. Odor can be fragrant to farinaceous, and taste is mild to unpleasant. Edible if the taste is mild, but cook well. Range NA, EU.

↓197d. Pale pinkish buff or yellowish buff spores; cap 1½–5 in. wide

Varieties of *Collybia* (= *Lepista*) *irina*

Collybia irina var. *luteospora*   *Collybia irina* var. *irina*

Margin is initially incurved. Variety *irina* (NA, EU, Asia, North Africa) has a pale pinkish buff spore print, while variety *luteospora* (NA) has a yellowish buff spore print. Fruits in boggy areas, late August through October. Uncommon. Odor and taste are pleasant, but many reports of GI distress after eating. Was long placed in the genus *Lepista*.

197e. Vinaceous buff spore deposit; cap 1–6 in. wide

Lepista densifolia

Dry cap, convex when young but margin uplifting in age, becoming flat or depressed in center. Mixed woods in the fall (Cascadia, EU). Mild odor and taste. Edibility? Difficult to ID correctly. Like all current and former *Lepista* species, was once placed in genus *Clitocybe*.

198a. (196a) Grows on the ground *200a*

198b. Grows on wood (appears terrestrial if on buried wood) *199a*

199a. (198b) Dark gray-brown cap; gills off-white to grayish, ± forked

Pseudoclitocybe cyathiformis

Cap 1–3 in. wide, pale, dry. On rotten hardwood or conifer logs and stumps, buried wood, in fall to winter. Odor and taste indistinct. Edibility? Formerly *Clitocybe cyathiformis* and *Omphalina cyathiformis*. *P. expallens* has a brown cap. Distribution NA, EU, Asia, North Africa.

↓199b. Smoky to dark blackish brown cap, 1–4 in. wide; gills white

Gerronema atrialbum

Cap initially convex, but margin uplifts in age to funnel shape. On wood, often alder or oak but also other hardwoods, in spring through early winter. No distinct odor or taste. Edibility not reported but not worth testing. Formerly known as *Clitocybe atrialba* and later *Clitocybula atrialba*.

199c. Olive-brown to grayish or blackish brown cap; gills white

Ampulloclitocybe avellaneialba

Cap (2–8 in. wide) is initially ± flat with a small umbo, margin upturns in age to funnel shaped. Stipe is brownish. On rotting alders and rotting conifer wood in the fall and winter in Cascadia. Indistinct odor and taste. Not edible. Formerly known as *Clitocybe avellaneialba*.

200a. (198a) Cap orange-brown, brownish, grayish, or blue *202a*

200b. Cap white to cream-colored *201a*

201a. (200b) Bright white cap, 1–6 in. wide; growing in dense clusters in sand and gravel along roads and trails, rarely in the woods

Leucocybe connata

Very distinctive species, often with a broad umbo. Gills are rarely decurrent. Fruits early summer through early fall at all elevations in NA, EU. Odor is mild, taste is sour. Edibility? Some reports of the toxin muscarine and a dog death after consuming. Formerly known by several names, including *Clitocybe dilatata*.

201b. Cap ½–1½ in. wide, creamy white with hoary covering when young

Leucocybe candicans

Formerly in genus *Clitocybe*. Cap initially convex, aging flat to shallowly depressed. On hardwood leaves, occasionally conifer needle beds, summer through fall NA, EU. Odor indistinct to herbaceous. Taste mild to bitter. Edibility?

202a. (200a) Grows in grass, cap (½–2 in. wide) whitish to dingy grayish buff

Collybia (= Clitocybe) 'rivulosa' (= C. dealbata, = C. sudorifica)

Very dangerously poisonous species loaded with the toxin muscarine (two look-alikes). Gills are closely spaced and adnate to decurrent. Often grows with the edible *Clitopilus cystidiatus/C. prunulus* (see key lead 320b [page 253]), which has a pinkish buff spore print (*C. rivulosa* print is white) and with the choice *Marasmius oreades* (see key lead 227b [page 202]), which has widely spaced, notched gills. Found in grasslands and open woods in NA, EU.

↓202b. Dense, white mycelial mat; gills close spaced; chalky

Leucopaxillus albissimus group   *Leucopaxillus gentianeus*

In NA, ± six varieties of *L. albissimus* known. Diameter varies, 2–16 in. White at first, aging to light tan. Odor varies from sweet to unpleasant, taste varies from mild to bitter. Possibly edible but difficult to digest. *L. gentianeus* varies from dull brown to reddish brown (three forms in NA, EU, Caucasus, North Africa). Odor is mild to farinaceous, and taste is bitter. Inedible. Both species grow under conifers and oaks in the fall, usually in deep litter.

↓202c. Caps ± 1–3 in. wide; spring and early summer, usually near snowbanks

'Clitocybe' glacialis   *'Clitocybe' albirhiza*

C. glacialis (to be renamed *Pseudolyophyllum glacialis*) has a hoary, grayish cap initially, but soon is tan, bleaching in the sun to almost white. Typically a cottony mass of white mycelium at the base of the stipe. Most often under spruce and true fir. Odor and taste are mild. Considered edible by some but not very tempting. *'C.' albirhiza* (to be renamed *Rhizocybe albirhiza*) is zonate and orange-tan when young but soon fades to off-white. Attached to the soil by copious, thick, white rhizomorphs. Under conifers in mountains of western NA. Odor and taste are disagreeable.

↓202d. **Odor distinctly of anise; cap < 2 in. wide, whitish to grayish brown**

Collybia (= *Clitocybe*) *deceptiva*

Collybia (= *Clitocybe*) *fragrans*

Both species are reddish brown with a striate margin when young and fresh, paler when dry. Some consider *C. fragrans* a synonym of *C. deceptiva*; others separate them based on spore color, white for *C. fragrans* and pale pinkish buff for *C. deceptiva*. Both grow under conifers in the fall. Edible, anise-like flavor, overpowering unless used in moderation. Range NA, EU, Asia.

↓202e. **Odor distinctly of anise; cap < 4 in. wide, bluish green**

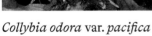

Collybia odora var. *pacifica*

Collybia odora var. *odora*

The two varieties differ only in gill color—white for variety *odora* and bluish green for variety *pacifica*. Variety *odora* is rare in Cascadia and variety *pacifica* is very common. Both are unmistakable. They fruit under hardwoods and conifers, summer through fall, at all elevations in NA, EU. Edible, with a very strong anise-like flavor. Was long placed in the genus *Clitocybe*.

↓202f. Grows in wooded areas; cap orange, brown, or pink (page 188) *206a*

202g. Grows in wooded areas; cap grayish to blackish brown *203a*

203a. (202g) Small to medium-size species; cap diameter 1–4 in. *205a*

203b. Large species; cap diameter averages > 4 in. *204a*

204a. (203b) Cap flat, funnel-shaped in age; no ring on stipe; flesh fragile

Aspropaxillus giganteus *Aspropaxillus septentrionalis*

A. giganteus (formerly *Leucopaxillus giganteus*) is 2–16 in. wide and grows summer to fall in open woods in NA, EU. Odor pleasant to farinaceous, taste is mild to disagreeable. *A. septentrionalis* (formerly *Leucopaxillus septentrionalis*) is 4–9 in. wide and found under conifers in summer to fall in Cascadia. Distinguished by a pungent to spermatic odor and a disagreeable taste. Neither is a good edible.

204b. Cap (3–16 in.wide) convex; double ring on stipe; flesh very hard

Catathelasma evanescens *Catathelasma 'ventricosum'*

Two species in Cascadia are differentiated by odor, spore size, and somewhat by habitat. *C. 'ventricosum'* usually has little or no farinaceous to cucumber-like odor and taste, is found under true firs and spruce, and has relatively small spores (9.8–12.2 µm x 4.0–4.7µm). *C. evanescens* has a strong farinaceous or otherwise unpleasant odor and taste and grows under true firs, larch, Douglas fir, alder, and birch (spores 13.5–16.7 µm and 5.1–6.1 µm). Both nontoxic.

205a. (203a) In old burns; dark cap, depressed center

Myxomphalia maura

The cap (½–2 in. wide) fades with loss of moisture. Gills white to grayish. Stipe cartilaginous, brittle. Often in moss in old burns, spring through fall in NA, EU, Asia, North Africa. Odor and taste are mild to slightly farinaceous. Edible?

↓205b. Brownish cap (1–4 in. wide) with slight umbo; often grape bubblegum–like odor

Ampulloclitocybe clavipes

Formerly placed in genus *Clitocybe*, *A. clavipes* fruits in mixed woods from summer to fall in NA, EU. Cap funnel-shaped in age. Cap and stipe are light to dark brown. Stipe (< 3 in. long) slightly to distinctly club shaped. Edible. *A. subclavipes* lacks the bubblegum odor and is distinguished microscopically.

↓205c. Grayish cap (< 2 in. wide) with white margin and umbo

Cantharellula umbonata

The white flesh sometimes stains reddish. Gills stain red or yellow. In boggy ground in mountains, summer and fall, exclusively in hair-cap (*Polytrichum* spp.) moss. Odor mild, fragrant to cucumber-like, and taste is mild. It is edible, but similar species are not edible.

205d. Grayish cap with darker center, no umbo

Spodocybe trulliformis PNW01

Cap (½–2 in. wide) is initially ± flat, eventually funnel-shaped. It fruits at all elevations under hardwoods and conifers, spring through fall. Odor and taste are farinaceous. Edibility? Formerly in genera *Clitocybe* and *Infundibulicybe*. Range of *S. trulliformis* group NA, EU. The Cascadia species is an unnamed look-alike.

206a. (202f) Cap and gills with pink tones; odor of pink bubblegum

Aphroditeola 'olida'

Distinct from EU and eastern *A. olida*, and needs new name. The ½–2 in. wide cap starts out flat, ages to funnel-shaped. Gills are distinctively forked and cross-veined similar to those in genus *Hygrophoropsis*. Under mountain conifers, July through October. Taste is mild. Edibility?

↓206b. Gills orange; forking; cap (< 4 in. wide) pale orange to orange-brown

Hygrophoropsis aurantiaca group

Genetically, this is a gilled bolete. The gills are ± orange, the cap and stipe can be completely orange or orange covered with tiny dark brown scales. Usually under conifers but can be under hardwoods, from late summer to fall in NA, EU, Australia. Odor mild, mushroomy. Taste mild to unpleasant. Possibly poisonous. Five unnamed look-alikes in Cascadia.

↓206c. Orange-brown cap (< 3 in. wide); strong farinaceous odor and taste

Bonomyces 'sinopicus'

Formerly *Clitocybe sinopica*. Found in open areas along roads and paths, sometimes old burns, from spring through fall in NA, EU. Cascadia species is distinct and needs a new name. Cap starts out smooth and flat to convex, becoming funnel-shaped in age. *C. diatreta* (fall-winter, page 318) has a mild odor and taste. Both are inedible.

↓206d . Pale pinkish tan cap (1–3 in. wide); faint anise odor

Infundibulicybe gibba

Fruits primarily under hardwoods, sometimes conifers, from July to October in NA, EU. Cap pinkish tan to flesh-colored, sometimes browner in age, initially flat to slightly depressed and becoming funnel-shaped in age. Taste is not distinctive. Edible but difficult to identify correctly. Formerly known as *Clitocybe gibba*.

↓206e . Cap (1–3 in. wide) ± brown with cinnamon, finely scaly disk, spring to fall

Infundibulicybe 'squamulosa'

Two unnamed varieties/species in Cascadia. Formerly known as the look-alike *Clitocybe squamulosa* (NA, EU). Cap flat, funnel-shaped in age. Under conifers and alders. Odor indistinct to farinaceous, taste ± disagreeable. Not edible.

206f . Cap (1–4 in. wide) orange-brown to red-brown; spore print white to creamy yellow

Paralepista gilva

Formerly in genera *Clitocybe* and *Lepista*. Gills may darken in age. *P. inversa* (NA, EU, Asia) is a look-alike or possibly the same species. Under mixed conifers in fall (Cascadia, EU). Odor mild to sharply peppery, taste mild. Not recommended. Edible?

207a. (176g) Lacking a partial veil and thus no ring on stipe (page 194) *212a*

207b. With a partial veil that leaves a fleeting or permanent ring on stipe *208a*

208a. (207b) Medium-large species; cap smooth or slightly scaly (page 192) *211a*

↓208b. Medium-large; cap and stipe with shaggy scales *210*

208c. Small (cap < 3 in. wide); cap and stipe with granular scales *209a*

209a. (208c) Cap (1–2 in. wide) and stipe tawny to dark brown with a sheathing ring

Cystoderma carcharias var. *fallax*

The flaring sheathing ring is a very distinctive feature, as are the granular scales covering cap and stipe. Also known as *C. fallax*. Under conifers and hardwoods from spring to winter in NA, EU, Asia. Common. Odor ± green corn, taste mild to nauseating. Not edible.

↓209b. Cap (1–2 in.) may be reticulate; fragile ring; odor ± green corn or may be disagreeable (if not reticulate, see also 209b [opposite]).

Cystoderma amianthinum group

Partial veil leaves a fragile ring. Cap and stalk color can be white or yellowish. Cap ± radially wrinkled. Common summer to winter under conifers in NA, EU, Asia. Cascadia has several closely related species/varieties, PNW01-03 plus *Cystoderma jasonis*, which is darker with a brown tint to flesh and gills. Taste indistinct. Not edible.

209b. (continued) Cap not reticulate (1–3 in.); fragile ring; odor not distinctive to oily/farinaceous

Cystodermella granulosa   *Cystodermella ambrosii*

Cystodermella cinnabarina

Cystodermella species were separated from genus *Cystoderma* based on DNA. *C. granulosa* is noted for its variable color, ranging from cinnamon to tawny. *C. ambrosii is* creamy white. *C. cinnabarina* is separated from *C. granulosa* by red-orange coloration, sometimes oily/farinaceous odor, and microscopic features. In all three species, stipe color = cap color. Partial veil leaves a hanging margin on the cap. Under hardwoods and conifers, late summer through fall in NA, EU, Asia, North Africa. Inedible.

210. (208b) Cap (2–7 in.) light yellow to yellow-brown; stipe white

Floccularia   *Floccularia albolanaripes* (two images)
'luteovirens'

Compared to *F. 'luteovirens'*, *F. albolanaripes* is darker (often with a brown central disk on the cap) and a bit smaller, but the two are difficult to distinguish and may be color forms of *F. albolanaripes*. Found in mixed woods in the fall, rarely spring, in western NA. Indistinct odor and taste. Possibly poisonous (unconfirmed).

211a. (208a) Cap (1–4 in. wide) orange-brown to brown, tiny darker scales in center

Armillaria gallica

Several difficult to distinguish species of honey mushrooms can occur under and on trees (see also key lead 306b [page 245]). *A. gallica* is usually found under and on hardwoods (especially oaks) in NA, EU, North Africa. Thin veil. Odor and taste faint, caps are edible but some adverse GI effects.

↓211b. White overall with yellow-brown stains or scales on cap (2–10 in. wide) and stipe; ring membranous and persistent; stipe may taper to a point (no bulb)

Tricholoma murrillianum   *Allotropa virgata*

T. magnivelare (eastern NA) was long incorrectly used for our western NA species. *T. murrillianum* (white matsutake) has exceptionally dense flesh. When you cut the stipe, the knife makes a distinctive squeaking sound. Odor is like a combination of dirty gym socks and cinnamon. Develops below ground, often showing just a rounded crown through the soil. Pick by holding the cap and rocking in a circular fashion until the whole mushroom pulls loose. Common under pines along the coast and inland where soil has a significant volcanic ash layer. Under mixed conifers, especially Douglas fir, starting in late August above 3,000 ft. Choice. *A. virgata*, candy stripe, is an obligate parasite on the mycelium. Compare to the seriously poisonous *Amanita smithiana* (key lead 187b [page 170]).

↓211c. Cap (2–5 in. wide) and stipe white with cinnamon to vinaceous brown scales; membranous veil leaving remnants on the cap margin; blunt stipe base; dense flesh; odor spicy cinnamon to disgusting

Tricholoma dulciolens/T. 'caligatum'

May be two species. Flesh is dense and white. Ring is sheathing. Uncommon. Under conifers in late summer to snowfall. May have same odor as white matsutake or may be unpleasant. Some collections edible and good, others dreadful tasting. *T. dulciolens* is known from Cascadia, EU.

↓211d. Cap (< 6 in. wide) and stipe dull brownish with orange or reddish brown

Tricholoma focale   *Tricholoma badicephalum*

These two species were once lumped together as *Tricholoma zelleri*, which is now considered a synonym of *T. focale* (as is *T. robustum*). *T. badicephalum* (Cascadia, Japan) and *T. focale* (NA, EU, Asia) fruit in the same habitat and season as white matsutake and are good matsutake indicators. Odor and taste generally mild. Sometimes severe GI distress if eaten.

211e. Cap (2–6 in. wide) and stalk sordid white or grayish; inconspicuous membranous veil with remnants on cap margin; rotting potato odor; spring into summer

Tricholoma vernaticum

Found under mountain conifers in Cascadia. The odor, like rotting potatoes, is distinctive, and the taste is farinaceous to cucumber-like. Not tempting as an edible.

212a. (207a) Caps 1–5 in. wide, densely clustered, tan to gray

Lyophyllum decastes group (above and at right)

Caps have a slippery feel but are not viscid, the gills are white to straw-colored. Odor and taste are mild. Edible. Three or more species in NA, EU, Asia, and North Africa. Avoid *Leucocybe connata* (a white species, see key lead 201a [page 183]) and *Entoloma* species (flesh-colored spore print, see *E. lividoalbum*, key lead 321a [page 253]). Eat only from clumps of five or more to avoid toxic look-alikes.

212b. Not growing densely clustered *213a*

213a. (212b) Cap width typically = or > stipe length *214a*

213b. Cap ½–3 in wide, width typically < stipe length; stipe fibrous with recurved scales; gills widely spaced, thick and waxy looking

Laccaria amethysteo-occidentalis   *Laccaria bicolor*

Easy to recognize the genus *Laccaria*, but difficult to differentiate the seven different orange species. (See also *L. laccata* image at key lead 1a #4 [page 42].) Common and abundant in all woodland habitats from mountains to sea level, summer to winter. Odor and taste are pleasant. Distinctive edibles, but not choice. *Laccaria* species are in disturbed areas of NA, EU, Asia.

214a. (213a) Caps white to yellow-brown, ± black fibrils (page 198) *219a*

↓214b. Caps gray or mouse-colored (page 196) *218a*

214c. Caps brown to orange *215a*

215a. (214c) Under mountain conifers near snowbanks; dark brown cap

Melanoleuca angelesiana

Cap dark brown to dark copper-brown, stipe paler, whitish to grayish gills, spore print white to cream-colored. Near high elevation snowbanks in western NA. Indistinct odor, mild to disagreeable taste. Edibility?

↓215b. Near oaks; fall

Tricholoma cf. dryophilum

True *T. dryophilum* (CA) bruises orange and is genetically distinct from this unnamed Cascadia species associated with Oregon white oaks. Cap (2–6 in. wide) is slightly viscid. Odor and taste are farinaceous. Edibility untested, but most likely a toxic species.

↓215c. Cap dry; under conifers and in mixed woods; summer to fall (page 196) *217a*

215d. Cap viscid; in mixed woods; summer to winter *216a*

216a. (215d) Cap (2–6 in. wide); under poplars and cottonwoods

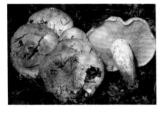

Tricholoma ammophilum

Distinct from *T. populinum* (EU). Gills and stipe develop reddish brown spots. Fruits (often in mass) in sandy areas at all elevations, summer to fall in NA. Odor and taste strongly farinaceous. Edible. Compare to *T. pessundatum* group.

↓216b. Cap (2–6 in. wide); mixed woods; scattered to densely massed

Tricholoma pessundatum group

T. pessundatum group includes six species, found in late summer through fall in northern NA. All bruise reddish. All are mildly to strongly farinaceous and inedible to poisonous. The very common *T. fulvum* is shown in the photo.

216c. Cap (2–4 in.) orange to red-brown; greenish orange scales on stipe

Tricholoma aurantium

Stipe has bands of greenish orange, scurfy flakes below a faint ring zone. Gills develop rusty brown stains. In mixed woods, summer to fall, in NA. Odor and taste strongly farinaceous. Unpalatable.

217a. (215c) Cap (2–8 in. wide) with matted ± scaly fibrils; gills brown spotted

Tricholoma 'imbricatum'

Robust species found under spruce and pines from late summer through fall. Genetically distinct from *T. imbricatum* (EU) and needs a new name. Odor and taste are mild to slightly farinaceous. Edible, but in dry weather easily confused with members of the toxic *T. pessundatum* group.

217b. Cap (2–4 in. wide) buff colored, with radial tan to dark red-brown fibrils

Tricholoma vaccinum

Gills buff, spotting then aging brownish orange; under spruce and pine summer to fall in northern NA. Odor and taste indistinct to farinaceous. Reportedly mildly poisonous.

218a. (214b) Cap (1–2 in. wide) scaly, grayish brown to black

Tricholoma moseri group

Indistinct, fleeting, cobwebby veil. Young margin inrolled and bearded with white fibrils. Cap not umbonate. Under conifers near melting snowbanks in western NA mountains. Odor and taste mild to farinaceous. Edibility? Group probably includes *T. triste*, *T. bonii*, and others.

↓218b. Cap (1–2 in. wide) with very black scales; stipe scaly

Tricholoma 'squarrulosum'

Member of the *T. atrosquamosum* group, each with an umbonate cap, a silky stipe, and a distinct but fleeting, cobwebby veil. Under conifers in fall in NA, EU. Indistinct odor and taste. One of ± twelve very similar *Tricholoma* species. Edibility untested, poisonous look-alikes.

↓218c. Dry cap (2–8 in. wide) white with small gray to tan, spot-like scales

Tricholoma 'pardinum'

Species shown here is unnamed but is like *T. pardinum* (EU). Fleshier, larger, and paler than other dry, gray *Tricholoma* species. Under hardwoods and conifers, in all elevations in NA. Odor and taste mild to farinaceous. Causes severe GI distress if eaten. If the scales are tan, it is *T. venenatoides*.

↓218d. Dry cap (2–4 in. wide), grayish, brown, or purple; pointed umbo

Tricholoma subacutum

Cap has a streaked look from radially arranged fibrils and may split radially. Under hardwoods and conifers at all elevations, late summer through fall in NA. Odor mild to musty, taste peppery to bitter. Possibly toxic. Long known as either *T. virgatum* or *T. argenteum*.

218e. Viscid cap (2–5 in. wide), grayish, brown, or purple; broad umbo

Tricholoma portentosum group

Viscid cap, in age flat to broadly umbonate. Radially arranged fibrils may split radially. Mixed woods, fall in NA, EU, Asia, North Africa. Odor and taste are mild to farinaceous. *T. portentosum* is edible but does not appear to be in Cascadia, which has *T. griseoviolaceum*, maybe one or two others.

219a. (214a) Odor and taste mild to farinaceous *221a*

219b. Odor strongly floral to repulsive; taste nauseating *220a*

220a. (219b) Dry cap (2–5 in. wide) and stipe; whitish with orange-yellow patches

Tricholoma lutescens

Thick-fleshed, hemispheric cap when young, flat in age. Under hardwoods and conifers in late summer to fall in Cascadia. Odor fruity to stinking sulphurous. Taste nutty to slightly peppery. Edibility? Nauseating. Long called *T. sulphurescens*, EU look-alike.

↓220b. Dry, slightly umbonate cap, 1–2 in. wide; overall white to pale tan

Tricholoma inamoenum

Gills distant. Mainly in montane habitats under spruce, summer and fall (NA, EU). Odor strongly floral to sulphurous, taste disagreeable. Unpalatable. *T. platyphyllum* is very similar (WA to CA and possibly east to CO).

220c. Dry cap (1–3 in. wide) and stipe; overall yellowish

Tricholoma 'sulphureum'

Overall yellow color, sometimes tan on the disk. Found at all elevations under hardwoods and conifers, late summer to winter. Odor is strongly floral to repulsive and sulphurous, taste is similar. Unpalatable. May be two species in Cascadia: *T. 'sulphureum'* (slender, NA, EU) and *T. odorum* (stocky, NA).

221a. (219a) Cap viscid *222a*

221b. Dry cap (2–6 in. wide) with a slippery feel, multicolored, ± yellow-green

Tricholoma saponaceum group

Commonly encountered under both hardwoods and conifers, spring to fall in NA, Sweden. Can be gray, resembling *T. virgatum*, or yellow, yellow-green, or even brown. Possible splash of pink at very bottom of stipe (dig carefully). Odor and taste soapy. Four species known so far. All are inedible.

222a. (221a) Cap (1–4 in. wide) slightly viscid with blackish radial fibrils over yellow

Tricholoma atrofibrillosum

Cap splitting radially, overall dark greenish yellow, paler stipe. Under hardwoods and conifers, late summer to fall. Odor indistinct to farinaceous. Taste mild to nauseating. Insipid to possibly poisonous. One distinct and unnamed species in Cascadia. *T. sejunctum* is a NA look-alike not in Cascadia.

222b. Cap (2–8 in. wide) yellow to brown; yellow gills; white to yellow stipe

Tricholoma yatesii   *Tricholoma equestre* group

Usually under pines in sandy soil, fall. Odor not distinct, of coconuts or mildly farinaceous. Taste mild to farinaceous. Three or more species in Cascadia do not include *T. equestre* (= *T. flavovirens*), which has caused deaths in Europe. Image at right is a conifer associate close to *T. ulvinenii* (EU), while *T. yatesii* is an oak associate. *T. intermedium* (page 318) is a white-gilled member of the group. Edibility?

223a. (176f) Mushroom grows singly or in small clusters; not cespitose *225a*

223b. Mushroom grows in tight clusters of > half-dozen caps *224a*

224a. (223b) Cap (< 1½ in. wide) ± convex, dark reddish brown; gills brown

Gymnopus fuscopurpureus

Cap hygrophanous, deep green with KOH. Gills brown when fresh, duller in age. Stipe ± cap color, tomentose below. Coniferous forests in fall, in western NA, EU. Odor and taste mild. Edibility? *G. putillus* (whitish stipe, NA) does not turn green in KOH. (See also the *Gymnopus spongiosus* group [page 318].)

↓224b. Tightly clustered; purplish red cap, < 2 in. wide; on rotten conifer wood

Connopus (= Gymnopus) 'acervatus'   Gymnopus erythropus

C. bagleyensis n.p. is the tentative name for *C. 'acervatus'*, the Cascadia look-alike to *C. acervatus* (EU). *G. erythropus* (NA) has a bicolored stipe and cream-colored gills, KOH turns the cap green. Both found from summer to fall, with mild odor and taste. Not edible.

224c. Cap (< 2 in. wide) ± convex, reddish brown fading; gills pinkish buff

Collybiopsis confluens

Tough, pliant stipe is ± cap color and covered with minute white hairs. On the ground in hardwood and coniferous forests, summer to fall, in NA, EU, North Africa, Asia. Odor and taste mild. Technically edible but tough. Formerly placed in genera *Gymnopus* and *Marasmiellus*.

225a. (223a) Stipe diameter ≤ soda straw (page 204) *231a*

225b. Stipe diameter > soda straw *226a*

226a. (225b) Cap not purple-red to pinkish, gills whitish (page 202) *227a*

↓226b. Gills bright yellow; purple-red cap, 2–4 in. wide; pinkish stipe

Calocybe onychina

Distinctive. Under conifers in summer in NA (rare). Odor and taste mild to farinaceous. Probably edible but infrequently seen and untested.

226c. Gills white; pinkish cap, 2–4 in. wide; pinkish stipe

Calocybe carnea

Pinkish cap and stipe, crowded white gills. Odor none. Taste indistinct. Edible, not recommended. Widespread in NA, summer–fall in grass and open places, bogs.

227a. (226a) Gills not widely spaced; stipe fleshy and breaks ± easily *228a*

227b. Gills widely spaced; stipe tough and pliant; hygrophanous; in grass

Marasmius oreades

Cap (½–3 in. wide) reddish tan, fading to tan. Dried specimens revive when moistened. Often found in rings; mycelium can live for hundreds of years. Mild odor with hint of apple seeds (an organic cyanide), mild taste. Caps edible. Found in NA, EU. Toxic *Collybia rivulosa* (key lead 202a [page 183]) has close, decurrent gills.

228a. (227a) Stipe not striate, ± longer than the cap diameter *230a*

228b. Squat; stipe striate, ≤ cap diameter (*Melanoleuca* spp.) *229a*

229a. (228a) Cap (2–5 in. wide) convex to broadly umbonate, brown, fading to tan in age (see key lead 215a [page 195] for additional *Melanoleuca* species)

Melanoleuca cognata

Cap smooth, slightly viscid. Spore print creamy or yellowish, stipe cap color or paler. Under conifers and in mixed woods, spring to fall in Cascadia. Odor and taste sweetish. Edible. Possibly several species.

↓229b. Cap (2–7 in. wide) broadly convex, pallid grayish brown, aging slightly ochre

Melanoleuca exscissa

Was long knolwn as *M. evenosa*. Common in grass in lowland areas in spring and early summer in Cascadia, EU. Distinguished from *Tricholoma* spp. by smooth cap and crowded gills. Odor and taste pleasant. Edibility?

229c. Cap (1–3 in. wide) convex to broadly umbonate, dark brown to gray, fading

Melanoleuca melaleuca group

Distinguished from *M. cognata* complex by a white spore print and microscopic features. Odor mild. Edible. In open places in NA, EU. Distinguished from *Tricholoma* spp. by smooth cap and crowded gills.

230a. (228a) Cap (2–4 in. wide) ± chestnut-brown; almond-like odor

Rhodocollybia oregonensis

Cap slightly viscid, gradually fading in age. Reddish stains on gills and stipe. Stipe 2–10 in. long, with a long underground portion. Fruits near old stumps and in lignin-rich humus, in fall in Cascadia. Odor strongly of almond, taste faint to bitter. Edibility?

↓230b. Cap (2–5 in. wide) pallid to pale tan, may bruise rusty red

Rhodocollybia maculata var. *occidentalis*

All parts bruise rusty red to reddish brown. Four or five known varieties. On much-decayed wood in coniferous forests, spring to winter, mainly fall in NA, EU. *R. maculata* var. *immutabilis* is more grayish and does not stain. Varieties vary from mild to bitter in taste.

↓230c. Cap (1–4 in. wide) with greasy feel, ± chestnut-brown; hygrophanous

Rhodocollybia butyracea

In humus under conifers, usually with pines, early spring into winter in NA, EU, North Africa, Asia. Spore print cream-colored to pinkish buff. Odor and taste mild. Mediocre edible. Gray- to brown-capped form, *R. butyracea* var. *asema*, dries almost white. May be additional closely related species.

230d. Cap (1–2 in. wide) with greasy feel, ± chestnut-brown; hygrophanous

Gymnopus dryophilus complex

Distinguished from *R. butyracea* by a white to pale yellowish spore print. In humus in conifer-hardwood forests, often with oaks in NA, EU. Odor and taste mild. Edible caps, but difficult to ID and some reports of GI distress after eating. There are seven known species in this complex.

231a. (225a) Growing on buried wood or on the ground *233a*

231b. Growing on decayed old (*Russula* spp.) mushrooms *232a*

232a. (231b) Cap (< ½ in. wide) flat, or growing from a sclerotium

Collybia tuberosa   *Collybia cookei*

C. tuberosa grows from an apple seed–like sclerotium (lower left, left photo). *C. cookei* differs in growing from a roundish, flat, yellow to yellow-orange sclerotium (lower left, right photo). *C. cirrhata* (page 319) is similar but has no sclerotium. All are found in summer and fall, often on dead *Russula* spp., in NA, EU, Asia.

232b. Cap (< 1 in. wide) ± hemispheric, silky white to gray; gills broad, distant

Asterophora parasitica   *Asterophora lycoperdoides*

Gills of *A. parasitica* are distant, broad, ± interveined, and spores drop normally. *A. lycoperdoides* has a few thick white to pale gray gills, but otherwise is like a puffball, with the white cap disintegrating to reveal a cinnamon-colored spore mass. Both grow primarily on rotting members of the *Russula nigricans* complex in the late summer and fall in NA, EU, Asia. Odor and taste are strongly farinaceous. Too tiny to explore the edibility.

233a. (231a) Stipe smooth, possibly with whitish tomentum at the base *235a*

233b. Stipe base or whole stipe covered in short, velvety brown hairs *234a*

234a. (233b) Cap (1–3 in. wide) ± broadly umbonate, tawny to red-brown

Flammulina velutipes group

Cap smooth and viscid. Usually velvety stipe darkens to dark brown. Found winter to spring. Odor and taste pleasant. Edible. Includes *F. lupinicola* (on lupine), *F. populicola* (usually on aspen), *F. filiformis* (illustrated, on hardwoods, found in western NA, EU), and *F. rossica* (often on willow).

234b. Cap (1–2 in. wide) bell-shaped, tawny to red-brown; cucumber-like smell

Macrocystidia cucumis group   *Macrocystidia cucumis* var. *leucospora*

Cap velvety, hygrophanous. Stipe cinnamon-colored, aging to dark red-brown. Rare in forests, common in grassy areas, spring to fall in rich soils in NA, EU, Asia, North Africa. Odor often fishy, cucumber-like, taste farinaceous. Not palatable. At least four different species in Cascadia.

235a. (233a) Cap convex to bell-shaped, maturing umbonate (page 206) *237a*

235b. Cap expanding to flat or with a central depression in age *236a*

236a. (235b) Cap (< ½ in. wide) ± flat, white to pinkish buff; on conifer cones

Strobilurus trullisatus/S. occidentalis

These two species, plus *S. albipilatus*, are separated microscopically. *S. trullisatus* grows on Douglas fir cones. *S. occidentalis* grows on many different cones. *S. albipilatus* (page 319) grows on cones and debris. All are common in summer to winter in NA. Indistinct odor and taste. None are edible.

↓236b. Cap (< 1½ in. wide) hygrophanous, flat, sunken center, red-brown

Collybiopsis villosipes

Cap wrinkled, thin, and leathery, fading to dull brown when dry. Gills appear free when old. Stipe tough and wiry. In duff under conifers and in wood chips, in disturbed areas in Cascadia. No data on odor, taste, or edibility. Formerly in genera *Gymnopus* and briefly *Marasmiellus*.

236c. Cap (< 1 in. wide) convex to flat, yellow-brown to orange-brown

Xeromphalina cauticinalis

Cap bald, moist. Gills crowded. Stipe reddish brown to black, apex yellowish, tomentose with orange tomentum at base. In conifer or alder debris in spring and fall in NA, EU, Asia. Odor pleasant, taste bitter. Inconsequential.

237a. (235a) Dried-out mushroom not reviving when moistened *239a*

237b. Dried-out mushroom reviving when moistened *238a*

238a. (237b) Cap (< 1½ in. wide) dry, reddish brown, darker on disk

Marasmius cohaerens var. *cohaerens*

Cap is tough and pliable, smooth; revives when moistened. Stipe is pliant to cartilaginous. Gills whitish, distant (gills close in variety *lachnophyllus*). In hardwood debris (sometimes with conifers), summer to winter in NA, EU. Odor earthy, taste a bit unpleasant. Edibility?

238b. Cap (< 2 in. wide) dry, velvety, ± wine-red; red-black, brittle stipe

Marasmius plicatulus

Cap is tough and pliable, obtusely conic to distinctly bell-shaped, velvety, and appearing frosted when fresh. Stipe base with extensive mycelial mat. Grassy areas and mixed woods, spring to winter in Cascadia. Moldy odor and taste. Too tough and thin to consider eating.

239a. (237a) Cap (< 2½ in. wide) broadly umbonate, purple, gray, bluish

Mycena pura

A very atypical and highly variable *Mycena* species. There is a yellow form, a white form with pinkish gills, a pink form with whitish gills, and it is often purple. All are large and wide (for a *Mycena* species). Most typical form is shown. Found spring to fall in all types of woods and all elevations in NA, EU, Asia, North Africa. Radish-like odor and taste, probably toxic, may contain muscarine. The color forms are genetically close; some include color varieties in a single species, though others consider them members of a group.

239b. Cap (< 1 in. wide) convex to flat, small umbo, brown fading to tan; on conifer cones

Baeospora myosura

Cap dry to soapy feeling, stipe tough and pliant, somewhat rooting. Gills very closely spaced. On cones of many different conifers, especially spruce and Douglas fir, from summer to winter in NA, EU. Faint mushroomy odor and taste, not palatable. Compare to *Strobilurus* species (key lead 236a [page 205]).

240a. (176e) Growing on ground or very rotten wood (page 210) *246a*

240b. Growing on wood *241a*

241a. (240b) Growing singly to gregarious, but not in dense clusters (page 208) *242a*

241b. In dense clusters on wood

 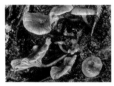

Xeromphalina campanella group

Cap (< 1 in. wide) varies from yellow-brown to orange-brown, develops central depression. Stipe ± yellow at apex, red-brown below, and covered in a tuft of short orange hairs at the base. On conifers. No distinct odor or taste, not edible. *X. enigmatica* differs only by DNA and appears to be a lowland species, while *X. campanella* is in the mountains. *X. brunneola* (right image) has a dull orange to reddish brown cap and is also found on rotting conifer logs. It has a disagreeable odor and taste. All three species fruit from summer until snowfall in NA, EU, Asia, North Africa.

242a. (241a) Gills brightly colored (yellow, orange, lilac) *244a*

242b. Gills white, grayish white, or pale orange-brown *243a*

243a. (242b) Stipe very slippery, covered in a clear slime layer

Roridomyces roridus

Formerly known as *Mycena rorida*. Dry cap (< ½ in. wide) is initially convex, later depressed, whitish to pale tan. Found spring through fall on coniferous sticks and needle litter in coniferous and mixed woods in NA, EU. Odor mild, taste not tested, too small and slimy to try and eat.

↓243b. Cuts ooze clear liquid; cap < 1 in. wide; stipe gray to black

Hydropus marginellus

Formerly known as *Mycena marginellus*. Convex cap becomes either slightly umbonate or depressed in age. Gill edges are brown to black. Clear liquid oozing from any cut is diagnostic. Stipe is initially minutely hairy, later bald. Found on conifer logs, spring to fall, in NA, EU. Mild odor and too small to try eating. Cascadia species is close to (EU) *H. marginellus* but possibly different enough that a new name will be required.

243c. Cap (< 1 in. wide) shining white, ± pinkish stains; widely spaced gills

'Marasmiellus' 'candidus'

Formerly *Marasmius candidus, Marasmius magnisporus, and Tetrapyrgos candidus*, this beautiful, distinctive species fruits in abundance on hardwood and conifer sticks and canes, from spring to winter in NA, EU, Asia, North Africa. Mild odor and taste. Cascadia may host two similar unnamed species. *Campanella* is the anticipated new genus.

244a. (242a) Cap vinaceous to cream-colored; young gills lilac

Chromosera 'cyanophylla'

Distinct from EU and eastern NA *C. cyanophylla* and is being renamed. Cap viscid, bald, striate. Lilac mycelium at stipe base. On rotten wood and debris, in wet coniferous woods, spring to fall. Indistinct odor and taste. Edibility?

↓244b. Cap < 2 in. wide; gills and stipe yellow to brownish yellow, fading

Chrysomphalina grossula (= *Omphalina grossula*)

Usually on conifer wood and debris, sometimes hardwoods, summer to fall in western NA, EU, Asia, North Africa. Waxy looking, similar to *Hygrocybe* species. Indistinct odor and taste. Edibility? *Chromosera citrinopallida* is lemon-yellow, fading to white, with the shape of *Chromosera 'cyanophylla'*.

244c. Gills light to dark orange *245a*

245a. (244c) Orange cap (< 1 in. wide), gills, and stipe; cap has small, whitish hairs

Chrysomphalina aurantiaca

Formerly known as *Omphalina luteicolor*. Cap, gills, and stipe fade to yellow on loss of moisture. Waxy look, similar to *Hygrocybe* species. On conifer wood and debris from spring to winter, mainly fall in NA. Indistinct odor and taste. Edibility?

245b. Cap (< 1½ in. wide) smoky orange-brown; gills ± orange-yellow

Chrysomphalina chrysophylla

Gerronema chrysophylla is one of several former names. Several color variants. Spore print colors range from yellow to salmon-buff. On rotten conifer logs and duff, spring to fall, in NA. Indistinct odor and taste.

246a. (240a) Stipes wiry-pliant, ± horsehair diameter *250a*

246b. Stipes ± diameter of a coffee stirrer/straw *247a*

247a. (246b) Cap light brown, pale yellow-brown to orange-brown *249a*

247b. Cap light brown, blue-green to blackish brown *248a*

248a. (247b) Cap (< 1 in. wide) dark blue-green to blackish green

Arrhenia chlorocyanea

Formerly known as *Omphalina chlorocyanea*, this distinctive species grows on open ground, in mosses, or on lichens in coniferous woods in Cascadia. Odor earthy to fishy. Taste mild. Edibility?

↓248b. Cap (< 1½ in. wide) hygrophanous, with pointed, fibrillose scales

Arrhenia gerardiana/A. bigelowii

Grows in sphagnum bogs, June to July, rarely August. Stipe long enough to project the cap above the moss. Formerly misnamed *A. sphagnicola*. Very similar sister species, *A. gerardiana* and *A. bigelowii*. Slightly farinaceous odor and taste. Edibility? *A. epichysium* is similarly colored but on wood and without the pointed scales. Distribution NA, EU.

↓248c. Cap (< 1½ in. wide) hygrophanous, no scales; gills medium to dark brown

Arrhenia telmatiaea

Found in sphagnum bogs and with other mosses in summer and fall, rarely June, in NA, EU. Was misnamed *A. onisca*. No distinctive odor or taste. Edibility? *A. rainierensis* looks identical but is found on soil or moss in wet road banks.

248d. Cap (< 1½ in. wide) hygrophanous, appressed scales, light brown; gills whitish

Arrhenia philonotis

Grrows in sphagnum bogs with *A. gerardiana* and *A. telmatiaea*. Gill edges and cap margin sooty brown, cap fades from pale brown to whitish. No distinct odor or taste. Edibility? Range NA, EU, with sphagnum.

249a. (247a) Cap (< 1½ in. wide) dull cinnamon color, fading to pale straw color, on green algae

Lichenomphalia umbellifera (± life-size)

Former names include *Omphalina ericetorum*. A globally distributed basidiolichen that is found at all elevations, spring to winter. On wood and always associated with the lichen *Botrydina vulgaris*. No noticeable odor or taste, inconsequential. *L. hudsoniana* is a hygrophanous orange-yellow species, circumpolar, in cold areas and in mountains.

↓249b. Cap (< ½ in. wide) bright orange, soon fading to orange-buff; whitish gills

Rickenella fibula (± life-size)

Formerly placed in genera *Omphalina*, *Mycena*, and *Gerronema*. Cap is striate, non-viscid, finely hairy when young. Found in moss in late spring through fall in NA, EU, Asia. Odor and taste are not distinctive, inconsequential. *Loreleia marchantiae* (Cascadia) is very similar, half the size, and is always on living or dead liverworts, summer to fall.

249c. Cap (< 1 in. wide) furrowed, depressed, orange; gills white to yellowish

Loreleia postii (± life-size)

Formerly placed in genera *Omphalina*, *Gerronema*, and *Clitocybe*. Found with mosses and liverworts near old burns or on wet soil around seeps, spring to fall in NA, New Zealand, Australia. No distinctive odor or taste. Edibility?

250a. (246a) Odor and taste like garlic or onions (page 212) *252*

250b. Odor mild, taste mild to bitter (page 212) *251a*

251a. (250b) Cap (< ¾ in. wide) ± amber; stipe dark; orange hairs at base

Both species start out convex and expand to flat with a depressed center. Both grow on conifer debris, spring to

Xeromphalina cornui   *Heimiomyces fulvipes*

fall. They differ in taste: *Heimiomyces* (= *Xeromphalina*) *fulvipes* (western NA) is bitter and *X. cornui* (NA, EU) is mild. Neither is edible.

251b. Cap (< ½ in. wide) dark reddish brown, fading to whitish

Pseudomarasmius pallidocephalus

Formerly in genus *Marasmius*. Cap dark reddish brown, fading to whitish. Hairless dark stipe inserted into hemlock, spruce, and Douglas fir needles, summer to fall in NA. Odor and taste mild, inconsequential. *Gymnopus androsaceus* (entire Northern Hemisphere, page 319) is very similar but prefers pine and redwoods.

252. (250a) Cap < 1 in. wide; on specific leaves or conifer needles

Mycetinis salalis   *Mycetinis scorodonius*

M. salalis (on salal and Oregon grape), *M. copelandii* (on oak and chinquapin), and *M. scorodonius* (on old fern fronds and conifer needle debris) all have the odor and taste of garlic. *M. copelandii* can fade to whitish when dry and can be smooth, wrinkled, or striate. *M. scorodonius* has a radially wrinkled cap, orange-brown to light brown, fading to reddish, and possibly is not quite as strongly flavored of garlic. All three species found spring to fall but are most common in the fall in Cascadia. *M. scorodonius* is known to be edible and has been used for garlic flavoring. All three recently in *Marasmius*. (See also *Gymnopus androsaceus* [page 319].)

253a. (176d) Mycenoid mushroom; growing on the ground (page 215) *257a*

253b. Mycenoid mushroom; growing on wood *254a*

254a. (253b) Cap yellow, reddish, purplish, blue (page 214) *256a*

254b. Cap buff to gray *255a*

255a. (254b) Odor radish-like, nitrous or bleach-like; cap (< 1½ in. wide)

Mycena cf. *leptocephala*

Complex of several species, on wood, sometimes growing in duff, from spring to winter. No data on taste. No edibles here. *M. silvae-nigrae* (Cascadia, EU) is dark brownish to dark grayish with an alkaline odor, common. *M. leptocephala* (probably illustrated here, global) is gray with alkaline odor. *M. abramsii* (CA only?) has a radish- or bleach-like odor and is gray to brown, darker in center.

↓255b. Cap (< 1½ in. wide) black-brown to yellow-brown; cespitose

Mycena 'maculata'

Often reddish spotted. Grows in large clusters on conifers or hardwoods in fall in NA, EU, North Africa. Odor and taste faintly spermatic. Not edible. Distinguished from *M. silvae-nigrae* by odor, tighter clusters. Small DNA differences may lead to a new Cascadia name.

↓255c. Cap (< 2 in. wide) dark blackish to pallid, greasy; near snowbanks

Mycena overholtsii

Stipe base often reddish and covered with white hairs. Fruits in clusters on old conifer logs, often right through the snow, in spring and early summer in Cascadia. Odor yeasty, taste unusual.

255d. Cap < ¼ in. wide; numerous on logs and living oak trees, fall to winter

Mycena meliigena (± 2x life-size)

Tiny species with a striate cap, distant broad gills, and ± cartilaginous flesh. On hardwoods, in NA, EU. Odor and taste are indistinct. *M. subcucullata* is distinguished by smaller spores.

256a. (254a) Cap < 1 in. wide, bluish to greenish aging to grayish or brownish

Mycena 'amicta'

Large and colorful when growing on conifer logs, smaller and duller colors when on conifer debris. It fruits summer and fall in NA, EU, Asia, North Africa. Cascadia species is genetically very different from (EU) *M. amicta* and a new name is required. Odor and taste are mild. Edibility? Blue color does not indicate psilocybin.

↓256b. Cap (< ½ in. wide) and stipe viscid, greenish yellow to bright yellow; on wood

Mycena epipterygia var. lignicola

Distinctive slippery yellow cap with white margin and yellow stipe. Odor is farinaceous. *M. epipterygia* var. *viscosa* has a gray cap and yellow stipe. *M. epipterygia* var. *epipterygia* has a yellowish gray cap, yellow stipe, grows on the ground. Neither are edible, both widespread NA, EU, Asia, North Africa.

↓256c. Cap (< 2 in. wide) reddish to pinkish brown; all parts exude red juice spontaneously or when cut

Mycena haematopus

Scalloped cap margin, coarse hairs on the base of the stipe, and cespitose growth habit. It is hygrophanous. Usually on hardwoods, spring to fall. An easily recognized global species. Odor fungoid, taste a little bitter. Too small to bother trying to eat.

256d. Cap (< 1 in. wide) dark purplish disk, paler margin; gill edges purplish

Mycena purpureofusca

Cap expands to nearly flat when old, fades a little on drying. Grows single to clustered on conifer wood, debris, and cones (NA, EU). No distinct odor or taste. Edibility?

257a. (253a) Gill edges the same color as gill faces (page 216) *259a*

257b. Gill edges brightly colored, darker than gill faces *258a*

258a. (257b) Cap (< ¾ in. wide) scarlet to yellow; gill edges scarlet to orange

Mycena strobilinoidea

Cap is scarlet when young, soon fading to orange and eventually yellow to whitish. Gills are yellow to pale pinkish orange and the edges are scarlet, fading to orange. Like many terrestrial *Mycena* species, it can fruit in huge troops (NA, EU). No distinctive odor or taste. Edibility?

↓258b. Cap (< ¾ in. wide) smoky olive-green, margin paler; gill edges orange

Mycena 'aurantiomarginata'

Cap feels greasy and is not hygrophanous. Gills are pallid to grayish olive-green with orange edges (need hand lens). Under conifers late in the fall season in NA, EU, North Africa. Indistinct odor and taste. Edibility? Cascadia species a bit different from the EU species.

↓258c. Cap (< 2 in. wide) with dark vinaceous brown center; gill edges vinaceous

Mycena elegantula

M. elegantula and *M. purpureofusca* appear to intergrade, and it is unclear exactly which species is present in Cascadia. *M. purpureofusca* (NA, EU)

has purplish gill edges, while *M. elegantula* (NA) has pale rosy gill edges and we observe intermediate colors between purple and pale rosy. The odor and taste are indistinct. Not edible.

258d. Cap (< 1 in. wide) vinaceous gray; bleach-like odor; gill edges rosy

Mycena capillaripes

Most of the gray to brown terrestrial *Mycena* species are tough to ID, but this species has a distinctive odor and distinctive gill edges. Densely gregarious on needle beds in summer and fall in NA, EU, Asia, North Africa. Odor is sometimes mild or of radish, but usually nitrous (like bleach), and the taste is slightly unpleasant. Edibility?

259a. (257a) Cap white with yellow center, brown, or brightly colored *261a*

259b. Mushroom entirely white, ± half the size of *Mycena* species *260a*

260a. (259b) Cap < ½ in. wide, sharp umbo or depressed; gills distant, decurrent

Hemimycena pseudocrispula

Initially the margin is depressed against the stipe and the cap extends beyond the gills. There is a very tiny basal disk where the stipe is attached to rhododendron leaves or conifer needles (OR, WA, EU). Summer to fall. The odor is indistinct, and the taste is a bit nauseating. Edibility?

↓260b. Cap < ¾ in. wide; strong nitrous (bleach-like) odor; decurrent gills

Atheniella delectabilis

Formerly *Hemimycena delectabilis*. Watery white when young, chalk-white in age. Distinct bleach-like odor. Fruits ± year-round on conifer needle beds in NA. Taste not distinctive, odor overpowering. Not edible.

260c. Cap < ½ in. wide; gills bluntly attached to free; odor and taste mild

Hemimycena lactea group

Grows on twigs and conifer needles, late summer to fall in NA, EU, North Africa, Asia. The close, narrow, ± free gills are distinctive. Inconsequential as an edible. Cascadia has at least four closely related species, some with and some without a bleach-like odor.

261a. (259a) Cap (< ¾ in. wide) viscid; yellow viscid stipe; near melting snowbanks

Mycena nivicola (= *M. griseoviridis var. cascadensis*)

This Cascadia species has a cap that is a variable combination of yellow, brown, and gray. In age, the disk is darker and the margin paler. The cap has a peelable outer skin. The odor and taste are strongly farinaceous. Edibility?

↓261b. Cap (< ¼ in. wide) coral-red, fading to yellow; stipe orangish to yellow

'Mycena' acicula (± 2x life-size)

Non-marginate gills, though edges may be white. Like *Mycena* species, but genetically distinct; at some point will be assigned a new genus. Grows on leaves and debris of hardwoods in especially wet areas, spring to fall in NA, EU, North Africa. The odor is mild, and the taste is indistinct. Too tiny to try eating it. Possibly two species present in Cascadia.

↓261c. Cap (< ¾ in. wide) orange when young, fading to yellowish white

Atheniella aurantiidisca

Can grow in large troops, carpeting the forest floor under Douglas fir and pines, spring and fall in Cascadia. Gills are white, tinged with yellow, and the edges are the same color as the gill faces. Odor and taste mild. Too small to consider for the dinner table.

↓261d. Cap (< 1 in. wide) creamy buff in center, white margin; whitish gills

Atheniella flavoalba

Formerly known as *Mycena flavoalba*. Cap is broadly to sharply conic, hygrophanous. The gill edges and gill faces are creamy white, gills not decurrent, at times interveined. On conifer needle beds and under oaks, late summer to fall in NA, EU, Algeria. Indistinct odor and taste. Not edible.

↓261e. Cap (< 1 in. wide) scarlet, fading to pink; pink gills with white edges

Atheniella 'adonis'

Formerly known as *Mycena adonis*. Distinct from the EU species and will need a new name. The cap is hygrophanous and can fade to white, and the gills fade to all white. In grass, moss, humus, on rotten wood in coniferous and deciduous forests in Cascadia, EU, Algeria. Indistinct odor and taste.

261f. Cap (< 1¼ in. wide) pinkish red, margin paler; gills flesh-pink or whitish edges

Mycena monticola

Cap is hygrophanous and can fade to flesh-color. Gills are bluntly attached with a small decurrent tooth. Found in coniferous forests above 3,000 ft. in Cascadia and can be gregarious to cespitose. Late summer to fall. Indistinct odor and taste.

262a. (176c) Waxy, viscid cap; dry stipe (page 222) *265a*

↓262b. Waxy, dry cap; dry stipe (page 221) *264a*

262c. Waxy, viscid cap; viscid stipe *263a*,
(see also page 225) *272d*

263a. (262c) Cap < 1 in. wide, initially green, bluish, or even orange, then many colors

Gliophorus psittacinus group

Once placed in genera *Hygrophorus* and later *Hygrocybe*, these distinctive species are widespread in woods and pastures, fruiting from spring to winter in NA, EU, North Africa. Cascadia has three known species but so far not *G. psittacinus* itself. The odor and taste are indistinct. Edible if you are not after flavor and do not mind slime.

↓263b. Cap (2–6 in. wide) olive-yellow to orange-yellow, darker center

Hygrophorus boyeri

H. hypothejus was a misapplied EU name. The gills start white and soon turn pale yellow, darkening in age. The stipe is slender (compared to *H. siccipes*), yellow to orange-yellow with ± reddish tones below viscid veil. Fall, under pines in NA. Indistinct odor, no taste.

↓263c. Cap (2–4 in. wide) dark brown (especially at center), ± red, orange

Hygrophorus siccipes

H. hypothejus was a misapplied EU name. Gills pale butter-yellow, later with orange and reddish tones. Common species under two- and three-needle pines, fall to spring in Cascadia. Inconspicuous fibrillose-glutinous veil. Indistinct odor and taste.

↓263d. Cap (1–3 in. wide) orange to orange-red; white to pale yellow gills

Hygrophorus speciosus var. speciosus

Cap margin fades in age to reddish yellow or golden yellow. Distinguished from red to orange hygrocybes by white decurrent gills. Stipe often club-shaped. Fleeting glutinous veil leaves slime on the stipe. Under larch or pine, in bogs, or woods in NA, EU. The odor and taste are mild. Edible.

↓263e. Cap < 1 in. wide, conic, reddish orange to yellow, bruising black

Hygrocybe singeri group

Usually called H. conica (a EU species complex), members of this species complex are distinguished from other similarly colored conic Hygrocybe species by bruising black in all parts. Common on the ground in many habitats in NA, Argentina. The odor and taste are indistinct. Probably poisonous.

↓263f. Cap (2–4 in. wide) white or yellowish with golden flakes on margin

Hygrophorus 'chrysodon'

Distinct from EU species and needs a new name. The delicate, golden granules can cover most of the cap. Fruits in the summer above 3,500 ft., fall and early winter in the low woodlands, winter in CA. Odor variable, not strong, taste mild to ± bitter. A bland edible.

263g. Cap 1–4 in. wide; gills and stipe pure white, aging slightly yellowish

Hygrophorus eburneus

Unclear whether Cascadia species is same as EU species or a slightly different species. The partial veil leaves slime on the stipe. Common under lowland oaks and in grassy areas in late fall in NA, EU. Late summer and fall, winter in CA. Faint, pleasant odor and taste. Edible but slimy.

264a. (262b) Cap < 2 in. wide; gills and stipe white, aging ± yellowish

Cuphophyllus virgineus   *Cuphophyllus borealis*

C. virgineus (NA, EU, Asia) often has pinkish lavender stains on the lower stipe, and both cap and gills turn slightly yellow in age. *C. borealis* (NA, Jamaica) does not have a staining stipe and yellows a bit only on the center of the cap. Both species can be found in deciduous and coniferous woods in late summer to winter, or to spring in CA. *C. virgineus* has a slight odor, a mild taste, and is edible and considered good. *C. borealis* has a mild odor and taste but is a flavorless edible.

↓264b. Cap 2–4 in. wide; gills and stipe dry, cream-tinted blue-green

Collybia odora (infected)
(was *Hygrophorus caeruleus*)

Found in spring and early summer under conifers in Cascadia. It is unique and has a strong order that resembles rancid meal. The taste is initially mild but soon turns unpleasant. Unpalatable.

↓264c. Cap (2–4 in. wide) dull orange to tawny; gills and stipe paler

Cuphophyllus pratensis group

Several similar species distinguished by microscopy and odor. Found in moist forested areas and in grassy areas in NA, EU, Asia, Argentina. The odor and taste are mild. Edibility reports range from fair to excellent.

264d. Cap (< 1½ in. wide), dry, ± scaly, fades

Hygrocybe constans   *Hygrocybe substrangulata*

H. miniata is a EU species. *H. constans* exists in several forms in pastures, moss, and forest debris. *H. substrangulata* (EU, NA), *H. cantharellus* (global, page 319), and *H. parvula* PNW01 (page 319) are in mossy bogs. All species are found from summer to fall and have little odor or taste.

265a. (262a) Indistinct to mild odor *268a*

265b. Distinctive, often aromatic, odor *266a*

266a. (265b) Odor ± strong of anise or almond extract (benzaldehyde) *267a*

266b. Odor aromatic, not like anise or almond extract

Hygrophorus pacificus

Cap (1–3 in. wide) is russet to tawny, margin fading in age. Gills have a yellowish cast. Stipe slightly yellowish or white. Grows in dense clusters under spruce in late summer and fall (western NA, NS). Odor penetrating, aromatic. Taste mild. Edibility?

267a. (266a) Cap (< 6 in. wide) brown in center, white margins

Hygrophorus bakerensis

Cap center reddish brown, yellow-brown, or gray-brown. Gills and stipe white to pinkish buff. Common under conifers, late summer to winter in NA. Odor usually distinctly of almond extract. Taste mild. Bland edible, no almond flavor when cooked.

267b. Cap (1½–4 in. wide) gray or brown; odor of almond flavoring or anise

Hygrophorus agathosmoides/H. odoratus

These two species are distinguished microscopically. *H. odoratus* (NA) is probably more common in Cascadia. Cap margin of both incurved when young, ash-gray to brownish gray. The white gills can turn grayish in age. Summer to fall under conifers and in mixed woods (NA, EU, Turkey for *H. agathosmoides*). Most common under spruce. Odor strong, sometimes like anise. Taste bland.

268a. (265a) Cap bright red, bright orange, or yellow (page 228) *275a*

↓268b. Cap white, dull orange, pinkish orange, gray, or black (page 224) *271a*

268c. Cap with reddish brown, pink, or purplish areas *269a*

269a. (268c) Spring to fall; cap (2–6 in. wide) slightly viscid; faint fibrillose partial veil

Hygrophorus 'purpurascens' group

Cascadia has one spring species and a different fall species, but *H. purpurascens* (EU) is a spring snowbank species. The spring species is low growing and often buried. Both grow mainly in the mountains under spruce and fir and have a mild to slightly anise-like odor and a pleasant to bitter taste. Edible.

269b. Lacking a partial veil; fruits fall to winter *270a*

270a. (269b) Cap (2–5 in. wide) multicolored, very viscid

Hygrophorus amarus

The ± bluntly attached gills are pale yellow and become pink-spotted. Fruits under spruce and Douglas fir from late summer until snowfall (western NA, NS). The odor is slight, and the taste is bitter to nauseating. Inedible.

↓270b. Cap (1–3 in. wide) whitish to pinkish, vinaceous in center

Hygrophorus erubescens

Cap viscid to slimy. Gills are bluntly attached to decurrent, white but soon developing pinkish to reddish brown spots. Generally above 3,000 ft. under pine and spruce from late summer until snowfall (NA, EU). Indistinct odor, pleasant to bitter taste. Edibility?

↓270c. Cap (1½–5 in. wide) light pink, with wine-colored to purple-red streaks

Hygrophorus parvirussula

Cap initially viscid, soon dry. Under hardwoods or in mixed woods, normally near oaks, not a conifer associate. Gills bluntly attached to slightly decurrent, white, developing purplish red spots. Odor and taste mild. Known from Cascadia and China, similar to *H. russula* (EU). Edibility?

270d. Cap (1–3 in. wide) purple-red or red-brown, viscid but soon dry

Hygrophorus 'capreolarius'

Gills are bluntly attached, decurrent in age, at first pallid, cap color in age. Mainly under spruce in boggy areas, sometimes in mixed woods in NA, EU, Asia. Indistinct odor and taste. Not reported to be edible. DNA of Cascade species differs from that of EU species.

271a. (268b) Cap dull orange to pinkish orange (page 227) *274a*

↓271b. Cap white (page 226) *273a*

271c. Cap gray to brown or black; fruiting with snowbanks and later *272a*

272a. (271c) Cap (1–3 in. wide) slightly viscid, streaked, dark brownish gray

Hygrophorus camarophyllus

Cap top-shaped, margin initially downy, can have a streaked appearance; gills conspicuously interveined. Near pine and spruce near snowbanks, spring to fall in NA, EU. Odor indistinct or faintly sulphurous, taste mild. Rated edible.

↓272b. Cap 2–6 in. wide; gills and stipe initially white, soon gray

Hygrophorus 'marzuolus'

This is an unnamed Cascadia species that has been known by the name of a EU look-alike. Often found near melting snowbanks in the mountains at high elevations. Can persist into the summer. Indistinct odor and taste. Edible.

↓272c. Cap 1–3 in. wide; stipe dry with dark fibrillose scales

Hygrophorus inocybiformis

Distinctive dark gray cap with remnants of the fibrillose veil on margin. Stipe dry, with dark gray fibrils below the ring zone. Under spruce and balsam fir in Cascadia, northern EU. Widespread but not common. Indistinct odor and taste. Edibility?

272d. Cap (1–5 in. wide) and stipe viscid; gray to dark brown fibrillose scales

Hygrophorus fuscoalboides

Stipe is white above the ring zone, ashy gray below. Found associated with both firs and hardwoods, September to November in NA. Indistinct odor and taste. Undesirable as an edible. *H. olivaceoalbus* (NA, EU) is darker, and *H. whitei* (CA, WA) has a slenderer stipe.

273a. (271b) Cap 2–8 in. wide; gills and stipe white; veiled when young; near snow

Hygrophorus subalpinus

Robust with a thick, dry, bulbous stipe and a ± membranous low ring. *H. sordidus* is similar but has neither a bulbous base nor a ring. Under conifers spring NA (has an unnamed fall look-alike). Odor and taste mild. Edibility good to poor.

↓273b. Cap (2–8 in. wide) viscid; gills and stipe dry, white; no veil; under oaks

Hygrophorus sordidus

Cap margin inrolled, slightly velvety, disk ± yellow-buff. Gills ± decurrent, slightly yellowish when old. Upper stipe slightly powdery. Mainly with oaks, fall in NA. Indistinct odor and taste. Mediocre edible. *H. ponderatus* has a fleeting partial veil.

273c. Cap < 2 in. wide, disk creamy yellow to flesh-color, margin white; anise-like odor

Hygrophorus pusillus

Bluntly attached white gills become short decurrent in age. Stipe turns pale yellowish white on handling. Under conifers, fall to winter in NA. Odor indistinct to fruity aromatic or of anise. Taste mild. Edibility?

274a. (271a) Cap (< 2 in. wide) viscid, ± rosy pink; near snowbanks

Hygrophorus goetzii

Cap fades to creamy white in age. The stipe is dry, often hairy at the base. Found on Mount Hood (OR) and Mount Adams (WA) near the tree line under conifers, notably hemlock. The gills are pallid, creamy white, or ± cap color. The odor and taste are mild. Too thin-fleshed to consider eating. Edibility?

↓274b. Cap (1–4 in. wide) whitish to pinkish cinnamon; flesh pinkish buff

Hygrophorus karstenii

H. saxatilis is a NA synonym of this EU/Cascadia species. The gills are short decurrent and ± pinkish cinnamon-tan. Fruits summer and fall under conifers. Odor indistinct to faintly apricot-like, taste mild. Edibility?

274c. Cap (2–5 in. wide) pale tan to pinkish orange; flesh mostly white

Hygrophorus pudorinus group (above and at right)

Cap convex to bell-shaped, margin initially inrolled. Gills at most slightly decurrent, pallid to pale pinkish orange. In boggy areas, mostly with spruce and pine in NA, EU. Odor not distinct, fragrant, or unpleasant. Taste indistinct to turpentine-like. Edible if mild tasting, but not worth messing with. *H. pudorinus* var. *fragrans* (left photo) is the largest and most common member of the group.

275a. (268a) Cap (< ¾ in. wide) viscid, red, aging orange-red; stipe orange to yellow

Hygrocybe subminiata

Gills are decurrent, white or tinged pale orange. Fruits in the summer and fall in NA, Jamaica. Odor and taste not recorded. Edibility? *H. minutula* (NA, Asia) is distinguished by a viscid cap and yellow gills. Members of the *H. miniata* group (264d, page 222) have dry caps and bluntly attached gills.

↓275b. Cap (1–3 in. wide) convex to flat, yellow to orange; gills bluntly attached

Hygrocybe 'flavescens'

The Cascadia species needs a new name because it is distinct from (eastern NA) *H. flavescens*. Stipe is yellow to yellow-orange, base is whitish. Feels slightly greasy but is not viscid. Gills white to typically yellow. Under both conifers and hardwoods, especially in damp, mossy areas, mainly fall. Odor and taste indistinct. Edible?

275c. Cap (1–4 in. wide) sharply to bluntly conical, bright yellow to orange

Gills notched to free, yellow. Under hardwoods and conifers, spring to winter in NA. Odor and taste mild. Harmless.

Hygrocybe 'acutoconica' (multiple species?)

Note: Taste *Russula* and *Lactarius* species for a peppery or burning taste. Chew for up to 30 seconds to detect latent heat. Peppery species lose their heat on cooking but sometimes cause stomach upset. Do not eat blackening species.

276a. (176b) Cut exudes clear or colored latex or reveals reddish or orange flesh; stipe color ± same as cap and does not resemble piece of chalk, though it breaks crisply (species of *Lactarius* [page 237] 293a)

276b. Cut flesh does not exude latex; stipe whitish or with blush of color, looks/breaks like chalk; bruising may cause color change to brown, yellow, red, gray, or black (species of *Russula*) *277a*

277a. (276b) Flesh very dense and compact; stipe difficult to break (page 235) *290a*

↓277b. Stipe medium hardness to hard; fragrant or nauseating odor (page 234) *289a*

↓277c. Stipe medium hard; spore print white to orange; mild tasting (page 231) *282a*

277d. Stipe medium hard; spore print white to yellow; peppery taste *278a*

278a. (277d) Spore print white (page 230) *281a*

278b. Spore print cream-colored to yellow *279a*

279a. (278b) Cap red or reddish brown; gills cream-colored to yellow in age (page 230) *280*

279b. Cap (2–4 in. wide) purple, ± tan, green; gills ± white; ± geranium odor; ± peppery taste; yellow to cream-colored spores

Russula 'queletii'

Two ± mild to peppery species associated with pine fit the photo and description of *R. 'queletii'* (NA, EU, hardwoods). *R. pseudopelargonia* (Douglas fir, hemlock) rarely shows tan to green tones. *R. salishensis* (Douglas fir, hemlock) fades to expose yellow to olive-green tones. *R. 'pelargonia'* (EU), with striate cap with shades of purple, gray, and red, could be in all forest types. *R. 'violacea'* (EU), with greenish cap and violaceous tones, may be in deciduous woods. All fruit from summer to fall. Edible?

280. (279a) Cap (2–4 in. wide) bright red, ± viscid; ± pink-flushed stipe

Russula americana   *Russula rhodocephala*

R. americana is mainly with Douglas fir and hemlock, late summer to fall.
R. rhodocephala is a pine species. Both have a burning-hot taste, and both
usually have a mild odor, though at times R. rhodocephala has a distinct fruity
or geranium-like odor. Neither should be consumed. Distribution: Cascadia.

281a. (278a) Cap (1–4 in. wide) cream-colored to creamy yellow

Russula crenulata

Cap skin peels easily, ± viscid. Gills white, finely
notched (saw-toothed). Under conifers and oaks, late
summer to fall in Cascadia. Indistinct odor, taste very
peppery. R. cremoricolor does not peel as easily, gills
not saw-toothed.

↓281b. Cap (1–3 in. wide) white, tan, or pale yellow; stipe yellowing

Russula pallescens

Cap skin does not peel. Flesh is hard, firm, and
elastic. Stipe bruises yellow. Fruits under hardwoods,
late summer to fall in Cascadia. Indistinct odor, taste
very peppery. Not edible.

281c. Cap (1–4 in. wide) vivid red to pink, yellow, or white; gills white

Russula emetica

Cap color variable, from pure red to pure white
or mixed colors. In swampy woods, conifers, or
hardwoods, summer to fall in NA, EU. Odor mild,
taste very peppery. Edible? R. montana (western NA)
stipe may gray a bit. Other rare look-alikes exist.

282a. (277c) Stipes firm; often a pithy core; easily breaks *284a*

282b. Stipes soft and easily crushed, often almost hollow inside *283a*

283a. (282b) Cap (1–3 in. wide) viscid when wet, yellow; yellow-orange spores

Russula 'postiana'/R. 'lutea'

Long known as the EU species *R. lutea*. Cap can be yellow, apricot-colored, or white. Associated with hardwoods. The odors are mild to fruity, and the taste is mild. Both are edible but thin-fleshed. DNA of Cascadia collections usually close to (EU) *R. postiana*, but some collections are close to *R. lutea*.

283b. Cap (1–4 in. wide) purple, purplish red, grayish olive-green; white spores

Russula hypofragilis

Grows where true firs are present in Cascadia. A host of similar species can be found under hardwoods and conifers, late summer to fall. Indistinct odor and usually mild taste. Probably edible but thin-fleshed and not worth the bother. *R. phoenicea* is similar, under Douglas fir.

284a. (282a) Flesh not bruising or bruising yellow or brown (page 232) *286a*

284b. Flesh may stain pinkish, soon grayish, and then blackening *285a*

285a. (284b) Cap (2–6 in. wide) blood-red to bronze, ± green, or lilac; cuticle half peels

Russula 'vinososordida'

Two species in Cascadia—an unnamed look-alike to (EU) *R. vinososordida* and the rarer *R. decolorans* (northern NA, EU). Spore print ochre. Reportedly with birch, but under conifers in Cascadia. Mild odor, mild to bitter taste. Edibility?

285b. Cap (2–5 in. wide) variable, often purplish with yellow-green center

Russula occidentalis

Cap skin separable, color extremely variable. Flesh firm, turning reddish and then slowly gray to black when cut. Gills pale yellow, spore print cream-colored. Under conifers, summer to fall in Cascadia. Odor weakly fruity, taste mild. Edible.

286a. (284a) Under conifers or hardwoods; late summer to fall *287a*

286b. Under oaks; cap (1–3 in. wide) dingy white, yellow to reddish center

Russula basifurcata

Fruits under oaks, November to December in Cascadia, NY. Spore print cream-colored. Odor and taste mild. Edibility?

287a. (286a) Mild to pleasant odor and taste, no seafood odor *288a*

287b. Odor when warm or old similar to shrimp or crab; cap 2–7 in. wide

Russula xerampelina group (above and at right)

Stipe is firm and filled with pith, bruises yellow and then slowly brown. Cap skin peels halfway to center, bright lemon-yellow while covered in duff, with exposed colors variable in red, purple, greenish, tan, velvety black, or any combination of these. Pale yellow spore print. Distribution NA. Taste mild. *R. benwooii* rarely stains brown, no shrimp odor. Both edible.

288a. (287a) Cap (2–4 in. wide) shades of brownish green; bruises rusty

Russula 'heterophylla'

Cap skin difficult to peel. Crowded, narrow gills initially white, yellow in age. Spore print white. Under hardwoods and mixed woods, late summer to fall in NA. Odor indistinct, taste mild. Edible. *R. 'aeruginea'* is olive-green with a yellow spore print. Species in Cascadia unclear.

↓288b. Cap (2–6 in. wide) variegated (patterned) in many colors; gills flexible, not brittle

Russula 'cyanoxantha'

Two species in Cascadia are distinct from (EU) *R. cyanoxantha*. One proposed name is *R. malva* n.p. Species in photo is undescribed. Cap purple, pink, brown, olive-green, orangish, or whitish. Spore print white to pale yellow. Odor and taste mild to pleasant. Edible.

288c. Cap (2–8 in. wide) olivaceous when covered, not peelable, velvety

Russula 'olivacea' (above and at right)

The Cascadia species is close to (EU) *R. olivacea*. There are two similar species in CA. The red color on one stipe is from a drop of phenol. Cap skin often wrinkled circumferentially. Flesh does not bruise. Spore print egg-yellow. Under hardwoods and conifers, summer to fall. Odor faintly fruity, taste mildly nutty. Edible and good.

289a. (277b) Spore print white; odor ± coconut-like, taste peppery

Russula crassotunicata

Cap surface a thick, rubbery skin that peels halfway. White flesh stains brownish. Distinctive gaps in pith of stipe. Conifers and mixed woods, summer to fall (NA). Taste can be bitter or peppery. Not edible.

↓289b. Spore print orange-yellow; odor of maraschino cherries, almonds, or nauseating; cap 2–7 in. wide

Russula fragrantissima group

R. laurocerasi, R. foetens (both eastern NA), and *R. fragrantissima* (EU) are misapplied names. DNA confirms four species in Cascadia, and all but one, *R. amerorecondita* (NA), are unnamed. Stipe interior with cavities (see key lead 289a). Under hardwoods and conifers, fall. Taste ± nauseating.

289c. Spore print creamy yellow; odor spermatic, oily, or mild

Russula cf. *cerolens* (*Russula sororia* group; above and at right)

We know of four moderate-sized species (cap 2–5 in. wide) in the *R. sororia* group, including three unnamed species lacking the darker center of *R. cerolens* (CA to BC), but they are difficult to separate. Found summer to fall in woods and nearby grassy areas. *R. cerolens*, the most common member, has an oily, ± unpleasant odor and a ± peppery, oily taste. All presumed poisonous.

290a. (277a) All parts turn red and then black, or black ± directly (page 236) *292a*

290b. May have moderate brown bruising, but not blackening *291a*

291a. (290b) Cap (3–10 in. wide) ± white, initially depressed, later funnel-shaped

DNA revealed several species in Cascadia, none the true eastern NA *R. brevipes*. All are dry, white to buff, have a white to pale cream spore print, little or no odor, and a mild to slightly peppery taste. Variety *acrior* has a pale green color at the top of the stipe and on the lower gills and is consistently peppery. Widespread, summer to winter. The mild varieties are edible. Some people like the firm, granular texture and ability to absorb flavors in cooking.

Russula brevipes group

Russula brevipes var. *acrior*

291b. Cap (1–3 in. wide) ± white, depressed center; very peppery taste

Distinguished from *R. brevipes* by the much smaller size and very peppery taste. Sometimes discolors brownish when damaged. Found under conifers, late summer to winter in Cascadia, MI. Odor slight, taste intensely peppery. Not edible.

Russula cascadensis

292a. (290a) Hard, whitish cap (2–7 in. wide) and stipe blacken without reddening

Russula albonigra

Cap is pallid when young, then brown, and finally black. The skin does not peel. All parts turn brown and then black in age or when handled. The gills are close to crowded and the spore deposit is white. Odor indistinct or like musty wine barrels, taste mild to slowly peppery or menthol-like. Toxic? Distribution NA, EU.

↓292b. All parts bruise red, and then slowly black; gills widely spaced

Russula nigricans

Cap (2–8 in. wide) depressed in center, whitish, soon sooty gray-brown, finally all parts entirely black. In deciduous and coniferous woods, summer to fall in Cascadia, EU. Odor none to unpleasant earthy, taste mild to slowly peppery. Edibility? Do not chance it.

292c. All parts bruise red and then slowly black; gills closely spaced

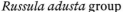

Russula adusta group   *Russula* cf. *acrifolia*

Long known as *R. densifolia*, a EU species. Slightly viscid cap (2–5 in. wide) is initially creamy white, soon shiny brown to smoky. Bruises red and then black overall. Late summer to fall in all forest types (*R. adusta* group, NA, EU). Odor faintly earthy, taste mild to peppery. If acrid, it might be *R. acrifolia* (EU). Poisonous?

293a. (276a) Milk white to a watery pale yellow-brown (like whey), not changing within minutes after cutting (page 240) *300a*

↓293b. Milk white, milk or flesh changes color within a few minutes of cutting (page 238) *295a*

293c. Milk can be orange to red (like raw beef), unchanging after cutting, always scanty, rarely forming any drops *294a*

294a. (293c) Cap flesh orange to orange-red; all parts may/may not turn green in age; *Lactarius deliciosus* group

Lactarius aestivus   *Lactarius vesper*

Lactarius aurantiosordidus   *Lactarius deliciosus* group

L. aestivus (most common) and the similar *L. vesper* exude latex ("milk") that turns more reddish after cutting (little or no green staining). *L. aurantiosordidus* also rarely stains green and is distinctively colored. At least two species like *L. deliciosus* (EU) stain distinctly green. Cap 2–6 in. wide, ± zonate. Edible, taste mild to bitter. Under hardwoods and conifers, late summer to fall in Cascadia.

294b. Cap flesh blood-red; all parts may turn green in age

Lactarius rubrilacteus

Cap (2-5 in. wide) zonate, redder than the *L. deliciosus* group. Under conifers and in mixed woods, late summer to winter (western NA). Odor indistinct, taste mild to slightly bitter. Edible but grainy and bland.

295a. (293b) Milk or flesh changing color or staining yellow to brown *297a*

295b. Milk changing color to ± lavender or flesh staining lavender after cutting *296a*

296a. (295b) Cap (2–8 in. wide) viscid, ± zoned, reddish brown, margin woolly

Lactarius repraesentaneus

Distinctive. Flesh slowly turns dull lilac or purplish where damaged. Milk color does not change. Under conifers, especially spruce, in NA, EU. Spores yellow in mass. Odor faintly fragrant, taste somewhat bitter and/or peppery. Poisonous?

296b. Cap (2–4 in. wide) color variable, white to grayish white, gray, brownish orange, or light brown with a grayish tint; slimy-viscid

Lactarius pallescens

In coniferous and mixed forests, late summer to fall in Cascadia, EU. Spores pale orange in mass. Odor indistinct, taste indistinct to slowly peppery. Possibly poisonous. *L. montanus* (page 38), *L. californiensis*, and *L. cordovaensis* (page 320) are similar purple-staining species.

297a. (295a) Cap margin bearded, coarsely hairy or woolly when young *298a*

297b. Cap (2–5 in. wide) margin not bearded, dry, orange zonate, milk turns yellow

Lactarius xanthogalactus

Cap color pinkish orange to orange-brown, ± zonate. Under hardwoods and conifers, late summer to winter (CA to WA). Odor indistinct, taste ± bitter peppery. Edibility? *L. chrysorheus* is an eastern NA species.

298a. (297a) Depressed, "wetted" spots on stipe, few to none *299a*

298b. Numerous distinct "wetted" spots on stipe; cap (1–4 in. wide) ages yellowish

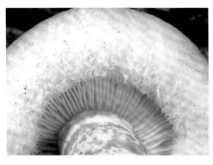 

Lactarius scrobiculatus var. *canadensis* (in *L. scrobiculatus* group)

Cap slightly viscid, white milk turns yellow in seconds. Spore print white to pale yellow. In mountains under conifers, late summer to early fall in NA. Odor faintly fragrant, taste not distinctive to slightly peppery. Not a recommended edible. *L. torminosus* var. *nordmanensis* (NA, EU) milk also yellows (see key lead 304c [page 244]).

299a. (298a) Cap (2–6 in. wide) initially white and azonate, viscid, bearded margin

Lactarius resimus group

Milk white, quickly changing to yellow. Spore print white to pale buff. Under mixed hardwoods and conifers, late summer to fall in NA. Indistinct odor, taste slightly bitter, ± slowly peppery. Edibility? At least two species in Cascadia.

299b. Cap (2–8 in. wide) yellowish orange, viscid, margin ± slightly hairy

Lactarius alnicola

Stipe at times scrobiculate. Cap ± zonate. Milk white, very slowly yellowing, and staining flesh yellow. Under alder and conifers, especially spruce, summer to fall in NA. Odor fungoid, taste instantly peppery. Not edible.

`300a.` (293a) Cap whitish, some lavender or gray (page 244) *304a*

`↓300b.` Cap viscid, medium brown to blackish (page 243) *303a*

`↓300c.` Cap viscid, yellow, tan, orange, or red-brown *302a*

`300d.` Cap dry, yellow, tan, orange-brown, or blackish *301a*

`301a.` (300d) Cap (1–4 in. wide) velvety dark brown; ± brown gill edges

Lactarius fallax group (above and at right)

L. fallax was once separated into two varieties, one with brown gill edges and the other with white gill edges. Two species are now recognized, and the two cannot be separated by color of the gill edges, a variable feature. Both grow with true firs in alpine habitats, late summer to fall in Cascadia. Yellowish spore print. Indistinct odor, mild to slightly peppery taste. Edibility?

`↓301b.` Cap (2–6 in. wide) ± umbonate, brownish red; very peppery taste

Lactarius rufus var. *rufus*

A common species usually associated with spruce in boggy areas, late summer to fall in NA, EU. Spore print ± cream-colored. Odor indistinct. Taste immediately or latently strongly peppery. Not edible.

`↓301c.` Cap (1–3 in. wide) red-brown, fading to pinkish brown; mild taste

Lactarius alpinus

Cap ± umbonate, vase-shaped in age. Milk almost clear, a pale yellow-brown. Spore print white. Under alder in wet areas, summer to fall in NA. Indistinct odor, taste mild. Edibility? Whether Cascadia has two varieties or two different species is unresolved. (See also *L. 'alpinus'* [page 320].)

301d. Cap (1–3 in. wide) rusty brown to orange-brown, with minute bumps

Lactarius rubidus (except for small, shiny specimen at far right)

The famous candy cap mushroom, also known as *L. fragilis*. The cap feels velvety. Fall to spring in many forest types in Cascadia. Common on the coast, southern OR to Santa Cruz, CA; rare elsewhere. Odor, if fresh, not distinct, powerful of maple syrup once dried (or if you hold a lighter to the cap margin). Taste mild, sweet when cooked. *L. subviscidus* is similar, slightly peppery; exudes thick, white milk; no maple syrup flavor. Non-toxic.

302a. (300c) Cap (1–3 in. wide) scarlet to orange, not zoned; slowly peppery taste

Lactarius substriatus

Has passed for *L. luculentus* and *L. aurantiacus*. *L. subflammeus* DNA is essentially identical, and *L. substriatus* is the older name. The cap is viscid. Stipe long relative to cap diameter. White milk exuded after cutting does not change color. Spores cream to white. Under pines, spruce, and other conifers in Cascadia. Indistinct odor, taste slowly peppery. Not edible.

↓302b. Cap (1–3 in. wide) reddish cinnamon color, not zoned; bitter or peppery taste

Lactarius luculentus (unnamed variety)

White milk may slowly stain gills brown. Cap ± viscid (variety *luculentus* is slimy-viscid). White spore print. Under alder and conifers in the mountains, late summer to fall in Cascadia. Indistinct odor, taste slowly bitter or mild. Not edible.

↓302c. Cap (2–7 in. wide) pallid to cinnamon-buff, vinaceous tones

Lactarius affinis

Cap not zoned, slimy-viscid, shiny when dry. Stipe color same as the cap. Milk white, not staining. Under conifers and in mixed woods and meadows, summer to fall in NA. Spore print white to yellowish. Odor mild, taste very peppery. Not edible.

↓302d. Cap (2–7 in. wide) slimy-viscid, cinnamon-tan, downy margin at first

Lactarius trivialis

Cap with smoky violet tones when young, obscurely zonate. Stipe pale cinnamon-colored to pale tan and color same as gills. Milk white, may slowly stain gills slightly olive-brown. Dull yellow spore print. Odor indistinct, taste slowly slightly peppery. Edibility? Range NA, EU. *L. cordovaensis* (= *cascadensis*) (Cascadia, page 320) wounds stain dull purple. Edibility?

302e. Cap (2–5 in. wide) viscid, distinctly zoned orange and apricot-colored

Lactarius olympianus

White milk does not change color after cutting, may stain flesh gray or olivaceous. Common in mountainous areas under conifers, summer to fall in western NA. Odor indistinct, taste extremely peppery. Not edible. When viewed from the top, *L. olympianus* looks much like members of the *L. deliciosus* group (see key lead 294a [page 237]), but they have orange gills and an orange stipe. Also compare *L. rubrilacteus* (see key lead 294b [page 237]), which has reddish orange cap, gills, and stipe.

303a. (300b) Cap (2–6 in. wide) slimy, blackish brown; white to pallid gill and stipe color; white milk slowly stains tissues darker

Lactarius kauffmanii var. *kauffmanii*

Milk slowly stains tissues olive-brown to gray-brown. Stipe slimy-viscid; gills and stipe pinkish buff to tan and stains brown. Common in coniferous woods, summer to fall in NA. Whitish spores. Indistinct odor, slowly peppery taste. Inedible.

↓303b. Cap (1–4 in. wide) slimy-viscid, dark smoky gray; white milk dries yellowish

Lactarius mucidus var. *fuscogriseus*

Gills have grayish cinnamon-buff spots in age. Gills close, pale pinkish tan. Viscid stipe, color same as the gills. Spore print pale pinkish buff. Indistinct odor, taste mild to slightly peppery. Edibility? Too slimy to eat. Range NA.

↓303c. Cap (1–3½ in. wide) slimy-viscid, dark smoky gray; white milk dries greenish

Lactarius mucidus var. *mucidus*

Cap may or may not have a low, sharp umbo. Gills white, edges fringed, stained blue–greenish gray from milk. Spore print white. Odor mild and taste peppery. Not edible. Range NA. *L. atrobadius* (Cascadia, page 320) tastes mild and gills are stained reddish brown.

303d. Cap and stipe viscid, dark grayish brown; white milk stains grayish

Lactarius pseudomucidus

Cap (1–4 in. wide) varies from blackish brown to dark gray. Stipe is the same color, very slimy. Gills white to light gray, brownish in age. Spore print white to pale yellow. Indistinct odor, slowly peppery taste. Not edible. Found throughout Cascadia.

304a. (300a) Cap (2–5 in. wide) viscid if wet, quickly dries whitish, soon pinkish buff; margin cottony woolly; taste peppery

Lactarius pseudodeceptivus

This species has thick, hard flesh that slowly discolors to tan when cut. The abundant white milk stains the gills brown. Under conifers, late summer to fall in NA. The odor is indistinct, and the taste is distinctly peppery. Not edible.

↓304b. Cap (3–9 in. wide) viscid if wet, quickly dries; distinctive pinkish tan gills

Lactarius 'controversus'

Cap may stain shades of lavender. Associated with willows and cottonwoods in the mountains, summer to fall in NA, EU. Odor indistinct, taste latent peppery hot. Not edible. The Cascadia species' DNA differs from the type by about 1 percent.

304c. Cap (1–4 in. wide) never viscid, pinkish orange; coarsely hairy margin

Lactarius torminosus var. *nordmanensis*

The cap, gills, and stipe are all pinkish cinnamon-colored. The spore print is pale yellow. Under cottonwoods, willows, and hemlocks in the mountains of NA, EU. Indistinct odor, slowly peppery taste. Not edible.

Note: Some of the mushrooms with a fleshy stipe and growing on buried wood (that look terrestrial at times) have been covered already. (See key leads 199a–c [page 182] and 221a [page 199].) Also, mushrooms with a slim stipe and growing on wood have already been covered. (See key leads 224b [page 200], 234a [page 205], 241b–245b [pages 207–209], and 255a–256d [pages 213–214].)

305a. (176a) Species bluntly attached to wood or with a stubby stipe (page 248) *309a*

↓305b. Species with a distinct stipe; saw-toothed gill edges (page 247) *308a*

305c. Species with a distinct stipe; smooth gill edges *306a*

306a. (305c) Grows single to gregarious but not cespitose (page 246) *307a*

↓306b. Cespitose, caps (1–4 in. wide) honey-colored; ring on stipe

Armillaria solidipes (= *A. ostoyae*) ("correct" name in debate)

A. solidipes (= *A. ostoyae*), the world's largest organism, is in the Ochoco National Forest (OR) and is the size and age of the forest. The second largest organism is in Gifford Pinchot National Forest (WA). Tiny black scales concentrated in the middle of the cap are distinctive to most honey mushrooms, except for smooth-capped *A. mellea*. Common summer to winter in NA, EU. Fruits at all elevations. The odor and taste are indistinct. Caps edible if thoroughly cooked, but a significant number of people suffer GI distress, sometimes hallucinations, after eating. Six *Armillaria* species are known from Cascadia. (See also *A. gallica*, key lead 211a [page 192].)

306c. Grows cespitose to singly; odor ± farinaceous; on hardwoods

Hypsizygus tessulatus

Cap (2–6 in. wide) pinkish cream, smooth, with distinctive watery spots when fresh. Stipe white, smooth except for stiff hairs at base. Uncommon, fall in NA, EU. Odor mild to farinaceous. Taste not recorded. Considered edible.

307a. (306a) Cap (3–8 in. wide) velvety, brown; stipe ± eccentric, velvety dark brown

Tapinella atrotomentosa

This distinctive and common species is a bolete relative. Gills rub off easily. Pale brown spores. Spore print yellowish or brownish. Found singly or in groups, July to October at all elevations in NA. Odor slightly unpleasant, the taste is mild to bitter. Inedible.

↓307b. Cap (1–4 in. wide) yellow, ± black or red fibrillose scales

Tricholomopsis spp.

Tricholomopsis rutilans   ### *Tricholomopsis decora*

These two are the most common and most distinctive of nine yellow-capped *Tricholomopsis* species. When *T. rutilans* is older, dark red to purple-red scales are sparser and the yellow ground color of the cap shows clearly. Both species grow on coniferous wood and wood chips in NA, EU. The odor of both species is indistinct and the taste is mild to watery. None of the *Tricholomopsis* species are good edibles.

307c. Cap (2–8 in. wide) whitish, pale yellow-brown, or grayish; membranous veil

Pleurotus dryinus

Veil leaves a slight ring on stipe, often remnants on cap margin. On living oaks and other hardwoods, summer to winter in NA, EU. Thick, firm flesh. Fragrant odor and pleasant taste. Edible but tough. Consider pressure cooking.

308a. (305b) Cap (1–2 in. wide) brownish, asymmetric, ± umbilicate

Lentinellus micheneri

Found late summer to fall in mixed woods on hardwood debris, conifer wood, or the ground in NA, EU. The odor is weak to sharply peppery, taste mild, then peppery or very bitter. Unpalatable. Formerly known as *L. omphalodes.*

↓308b. Cap (1–3 in. wide) convex to flat, rubbery-pliant, pinkish to tan

Neolentinus kauffmanii

The tough cartilaginous consistency is distinctive. The gills are smooth when young, saw-toothed at maturity. Uncommon on spruce logs spring to fall in Cascadia, Japan. Odor fungoid and taste fungoid to peppery. Inedible. Formerly *Lentinus kauffmanii.*

308c. Cap (3–20 in. wide) whitish with distinctive large scales, tough flesh

Neolentinus ponderosus   *Neolentinus lepideus*

N. lepideus cap white to pale yellow, dry to slightly viscid. Gills bluntly attached to decurrent. Membranous veil disappears in age. On conifers, sometimes oaks, spring to fall in NA, EU. Odor fragrant or anise-like, taste ± mild. Good edible when very young and tender. *N. ponderosus* (Cascadia) is similar but lacks the membranous veil and can be 20 in. across. It is choice when tender enough (try pressure cooking). Formerly in the genus *Lentinus.*

309a. (305a) Stipes rudimentary to absent; gills with smooth edges *310a*

↓309b. Gills with saw-toothed edges; cap (1–4 in. wide) brown; often near snowbanks

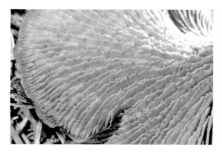 

Lentinellus vulpinus   *Lentinellus montanus*

L. vulpinus (EU, NA) is on hardwoods and *L. montanus* (NA) on conifers. In both, cap centers are coarsely hairy, odor slightly aromatic, taste slowly peppery. Thin, tough, inedible.

309c. Gills split in two lengthwise; cap (< 1½ in. wide) hairy, whitish; on hardwood

Schizophyllum commune group

The caps are fan-shaped, thin, tough, and leathery. The gills, when young, are whitish to grayish. The odor and taste are pleasant to sour. Too small and tough to be of interest. Several closely related, similar species share this name. Worldwide distribution.

310a. (309a) Gills white or very pale color *312a*

↓310b. Gills yellow to orange *311a*

310c. Gills silvery or pale tan, darkening in age; cap (< ¼ in. wide) dark

Hohenbuehelia unguicularis (± 2x life-size)

Fruits spring to winter on hardwoods in northern NA and EU, but rare or rarely noticed due to tiny size and dark color. Distinctive. Odor and taste not noted. Too tiny to eat. Very similar *Resupinatus applicatus* group members (global distribution) are distinguished microscopically.

311a. (310b) Entire mushroom dull to bright orange

Phyllotopsis nidulans

Cap (1–3 in. wide) densely furry, fading to tan in age. On well-rotted hardwoods and conifers, early summer to winter in NA, EU, Asia, North Africa. Odor mild and pleasant, or rather like rotten cabbage; taste mushroomy but unpleasant to bitter. Not tempting for the table. *P. nidulans* is in AK. The rest of Cascadia has a look-alike, *P. nidulans* PNW01.

311b. Cap (2–6 in. wide) multicolored in yellow, green, violet; gills pale yellow

Sarcomyxa serotina (= *Panellus serotinus*)

Late season usually on hardwoods, rarely conifers, in NA, EU. I have found it fresh and unfrozen at 14°F. Odor pleasant to indistinct, taste indistinct to bitter. A mediocre edible that is very tough. *Sarcomyxa* PNW01 (page 320) is known from *Quercus garryana* in Klickitat County, WA, CA. Edibility?

312a. (310a) Off-white, pinkish tan, or tan (page 250) *314a*

312b. Pure white *313a*

313a. (312b) Minute hanging or upright pitcher-shaped; lacking gills

Calyptella capula (± life-size)

Grows on wood or dead plant material in damp areas, summer to fall in NA, EU. Related to *Marasmius* species. Odor and taste not noteworthy. Easily mistaken for an ascomycete.

↓313b. Cap (< ¾ in. wide) dry, minutely hairy, not peelable; short lateral stipe

Cheimonophyllum candidissimum (± life-size)

Gills distant, yellowish in age. Fruits July to October on hardwoods and conifers in Cascadia. Indistinct odor and taste. *Panellus mitis* (Cascadia, EU) looks very similar but has a peelable cap due to a rubbery, gelatinous layer. Too tiny to eat.

313c. Cap (1–4 in. wide) smooth, may turn creamy in age

Pleurocybella porrigens PNW01

Angel wings, normally on hemlock logs (NA, EU, Asia), have a pleasant odor and unusual flavor. Long a popular edible. In Japan one year, however, several elderly people on dialysis who ate a huge quantity slowly died from holes in their brains. The species in Cascadia is genetically distinct.

314a. (312a) Cap < 1½ in. wide, pale ivory to pinkish tan or tan

S. longinquus (= *Panellus longinquus*) is hygrophanous, translucent when wet, often viscid. Mainly on red alder or on hemlock in late summer to fall in Cascadia, EU, Argentina, New

Scytinotus longinquus   *Panellus stipticus*

Zealand, Australia. Indistinct odor and taste. Edibility? *P. stipticus* is more brownish in all parts, very peppery, and weakly bioluminescent. Usually on hardwoods. Range NA, EU, Asia, Africa, Australia.

314b. Cap 2–10 in. wide, whitish, grayish, or bluish; on hardwoods

Three Cascadia species are in the *P. ostreatus* group. *P. populinus* is most common on cottonwoods, with buff-colored spores. *P. ostreatus* (page 320)

Pleurotus PNW07   *Pleurotus populinus*

is grayish to bluish with a lilac spore print. *P.* PNW07 is on alder and other hardwoods; spore print white to yellowish. *P. pulmonarius* (eastern EU) is a look-alike. All are edible with a pleasant odor and taste. Distribution NA, EU. (See also *Hohenbuehelia 'angustata'* PNW02 [formerly *H. petalloides*], page 320.)

315a. (175c) Gills attached; cap and stipe do not break like ball and socket (page 253) *319a*

315b. Gills free; cap separates from stipe in ball-and-socket fashion *316a*

316a. (315b) Small to medium-small; caps generally < 2½ in. wide (page 252) *318a*

316b. Medium to large; caps 2–8 in. wide *317a*

317a. (316b) Cap (2–5 in. wide) ± umbonate, dark brown to black; gill edges grayish black

Pluteus laricinus

The dark brown cap and smoky dark gill edges are diagnostic. Typically only on conifer logs and chips, spring to fall in NA. The odor and taste are usually pleasant. Soft flesh, but edible and pretty good. *P. laricinus* has long been known as *P. atromarginatus*, an EU look-alike.

↓317b. Cap (2–6 in. wide) white (but not cottony), gray to dark brown; stipe ± white

Pluteus exilis/P. cervinus

Look-alikes *P. cervinus* (NA, EU) and *P. exilis* are both found in Cascadia. Cap color and presence or absence of fibrils on stipe are highly variable. Normally on decayed hardwood logs, spring to fall. Positive ID requires microscopy and DNA. Odor and taste not distinctive. Edible, as are similar *Pluteus* species.

317c. Cap (2–8 in. wide) whitish to tan with dark ± black center; on hardwood

Pluteus petasatus

Large species closely related to *P. cervinus* group. Gills barely free, closely spaced. Fruits from spring until snowfall. Positive ID requires microscopy. Odor mild to radish-like, taste similar. A good edible. Global distribution.

318a. (316a) Cap (< 2½ in. wide) conic to broadly umbonate, cottony, white

Pluteus tomentosulus

Long known as *P. pellitus*. Found on hardwoods and conifers, summer to fall, often in swampy areas of NA. Indistinct odor and taste. Presumably edible (no toxic reports for *Pluteus* spp. except for hallucinations with some members of the *P. salicinus* group).

↓318b. Cap (< 2 in. wide) gray-brown to smoky, aging bluish; stipe whitish

Pluteus salicinus group

Except for *P. salicinus* itself, members of this group contain small amounts of psilocybin (with little effect). Usually on hardwoods (willow and alder), rarely conifers. See *P. cyanopus* in MycoMatch, a species with a brown, wrinkled cap. Both have a faint blue-green color on the stipe and are in NA, EU.

↓318c. Cap (< 2 in. wide) reddish brown, yellowish olive-green, brown

Pluteus fulvibadius (in P. romellii group)

P. lutescens is the former NA name. The presence of *P. romellii* (EU) or look-alikes is also possible. Cap typically granulose to velvety. Stipe base yellow. On hardwoods in fall. Indistinct odor and taste. Edible? There are ± twenty small *Pluteus* species in Cascadia.

318d. Cap (< 2 in. wide) umbonate, red to orange-red, fading to orange

Pluteus chrysophlebius

Cap often finely wrinkled. Stipe orange to red at base, the top lighter. Found on hardwoods, late summer to fall, often in small clusters. Odor indistinct, taste ± bitter. Edible but small. *P. aurantiorugosus* is distinguished by a mostly white stipe.

Note: The remainder of the pink-spored species are in the Entolomataceae, a family with about a thousand difficult-to-identify species, two edible species, and many species that cause severe GI distress when eaten. I include here common or distinctive species. I am treating *Nolanea, Leptonia, Entocybe, Claudopus,* and *Alboleptonia* as distinct genera, though many consider these subgenera of genus *Entoloma.*

319a. (315a) Cap ≤ 2 in. wide; stipe ≤ width of a soda straw (page 254) *322a*

↓319b. Cap small to large; stipe wider than a soda straw *320a*

319c. Cap ≤ 3 in. wide; stipe rudimentary or absent; ± on wood

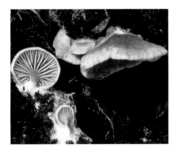

Claudopus byssisedus

Cap margin wavy, uplifted in age, covered by a whitish to grayish layer that disappears in age to reveal light brown to dark brown cap. Stipe ≤ ⅓ in. long. Fruits on old conifer or hardwood or on soil year-round, but uncommon in NA, EU, and Asia. Odor and taste farinaceous. Not a good edible.

320a. (319b) Gill attachment various but not distinctly decurrent *321a*

320b. Cap (2–4 in. wide) ± flat, slightly felty, white to gray; gills decurrent

Clitopilus cystidiatus/C. prunulus group

C. cystidiatus is common, *C. prunulus* (NA, EU, Asia, and North Africa) is probably present. Cap wavy, smooth, ± dry. In coniferous and deciduous woods, summer and fall. Odor and taste of bread dough. Edible, but resembles the seriously poisonous, white-spored *Collybia rivulosa* (see 202a [page 183]).

321a. (320a) Cap (2–6 in. wide) light to dark brown, bald; stipe whitish, stuffed and then hollow

Entoloma lividoalbum f. *lividoalbum*

There are numerous similar large *Entoloma* species in NA, EU, Asia, North Africa. All are very toxic. Abundant under both oaks and conifers below 2,000 ft. in fall. While this has no odor or taste, look-alikes can be farinaceous.

↓321b. Cap (< 2 in. wide) dark blue to ± black; stipe dark blue to blue-black

Entocybe nitida

Formerly known as *Entoloma nitidum*. Cap is shiny and not striate. Gills are close and pallid until colored pinkish by spores. In mixed woods or pure conifers, summer to fall. Odor mild to slightly radish-like, taste mild. Edibility?

321c. Cap (2–8 in. wide) bluish gray; bluish gray stipe, yellowish at base

'Entoloma' medianox

This species is easily recognized and was long known as *E. madidum* (EU) but is genetically distinct. Sometimes the cap is light pink to rosy. Will at some point be moved to a new genus. Common under California bay laurel, tanoak/madrone and other forest settings, fall through midwinter. The odor is indistinct to farinaceous, and the taste is farinaceous. Edible but not widely consumed.

322a. (319a) Dull-looking tan to dark brown (*Nolanea* species) (page 256) *326a*

↓322b. Beautifully colored, often blue (*Leptonia* and *Entocybe* species) *321b, 323a*

322c. Bright white cap, < 2 in. wide

Alboleptonia 'sericella' var. lutescens'

This common unnamed mushroom in Cascadia is an *A. sericella* (NA, EU) look-alike. The smooth, dry cap is pure white, aging yellowish or pinkish. Under hardwoods and conifers, late spring to winter. Odor and taste ± pungent. Not edible.

323a. (322b) Gill edges colored differently from gill faces *325a*

323b. Gill edges and faces are the same color *324a*

324a. (323b) Cap (< 1 in. wide) ± depressed, dark blackish blue, aging grayish

Leptonia 'parva' (old specimens)

In old specimens, the stipe and center of the cap are still blackish blue. Found on forest humus, mostly in fall in NA. Indistinct odor and taste. The colorful, scaly capped species may be split off from the more than fifty inedible *Lepiota* species.

324b. Cap (< 1½ in. wide) yellow-green to brownish olive; stipe yellow-green

Leptonia incana

Distinctive. The stipe bruises blue. Often in grassy areas and damp, open parts of woods, summer until snowfall in NA, EU. Uncommon. Odor has been described as similar to rancid butter, taste mild to disagreeable. Not edible. Not hallucinogenic.

325a. (323a) Gill edge blue; cap (< 1½ in. wide) and stipe dark bluish black

Leptonia 'serrulata'

An undescribed species, like *L. serrulata* (NA, EU). In age, the cap and stipe fade to bluish gray and the gills turn pink from the spores. Fruits spring to fall in rich humus or on moss-covered logs in deciduous or coniferous woods as well as in grasslands. Slightly farinaceous odor and taste. Three other *Leptonia* species have blackish gill edges, and many others lack colored edges. None are edible. A new genus may be proposed for the colorful, slender, scaly capped *Leptonia* species.

325b. Gill edge dark grayish brown; cap (< 2 in. wide) dark violet-brown

Leptonia fuligineomarginata

Cap and stipe can also be dark reddish brown with smoky tones. Found on the ground in mixed woods in western NA. The odor and taste are mild to pungent. Not an edible. Like most *Leptonia* species, microscopy is required for positive ID.

326a. (322a) Cap (< 3 in. wide) umbonate, hygrophanous, yellow-brown

Nolanea holoconiota

A common spring and early summer species found under mountain conifers in western NA. Has a two-toned cap. The odor and taste are generally indistinct. *N. verna* group members (EU, western NA) are dark gray-brown, and a non-hygrophanous spring species. *N. staurospora* is a farinaceous fall species in EU, Cascadia. None are edible.

↓326b. Cap (< 1½ in. wide) ± bell-shaped, hygrophanous, yellow-brown

Nolanea minutostriata

Cap is flat to bell-shaped, two-toned, darkest on the disk with a lighter margin, striate. Mainly fruits near cedars and red alder or vine maple in Cascadia (fall). Odor and taste indistinct. Not edible.

↓326c. Cap (< 2 in. wide) ± minutely umbonate, hygrophanous, gray-brown

Nolanea sericea

Found in grassy areas and on open ground year-round, except during the heat of midsummer in Cascadia. Cap is initially very dark and opaque, but is soon lighter and translucent-striate. Strongly farinaceous odor and taste. Not edible.

326d. Cap (< 2 in. wide) ± broadly umbonate, almost black, fading to brown

Nolanea edulis var. *concentrica*

Found in grassy areas near oaks, late fall through early spring in Cascadia. Unlike *N. sericea* (EU, NA), the odor is at most faintly farinaceous and the taste is mild. Like all *Nolanea* species, it is not a good idea to eat these because they may be toxic.

327a. (175b) Spore print orangish brown; young caps typically have cobwebby veil covering gills; stipe slender or thick, equal or with a bulbous base; gills ± bluntly attached (*Cortinarius* and segregate genera, a huge group) (page 278) *360a*

↓327b. Spore print dull yellow-brown to gray-brown; ± cobwebby veil; cap dry to moist with radially arranged fibrils and often splitting radially; stipe diameter typically ≤ pencil; gills dull brown to gray-brown, often with white edge (*Inocybe* and segregate genera) (page 274) *353a*

↓327c. Spore print dull brown; ± cobwebby veil; cap viscid, ± smooth; gills ± clay color, often with white edge; stipe thin to medium, ± equal; often radish-like odor (*Hebeloma* species, a medium genus) (page 273) *352a*

↓327d. Spore print dull brown; veil membranous fibrillose or absent; caps smooth, ± viscid, often cracked; grass or wood (*Agrocybe* species (page 272) *351a*

↓327e. Spore print yellow-brown to rusty brown; mostly on wood; veil fibrillose to membranous or absent; cap smooth to scaly, silky, or viscid; stipe often scaly, often in cespitose clumps (*Pholiota* and allies) (page 266) *339a*

↓327f. Spore print rusty orange; on wood; ± ring on stipe; caps ± silky, medium to large; bitter; often cespitose (*Gymnopilus* species) (page 263) *336a*

↓327g. Spore print various browns; stipe slender and ± fragile, ± ring (page 260) *331a*

327h. None of the above (small to large species from small genera) *328a*

328a. (327h) On wood; stipe absent or rudimentary; light brown gills rub off

Tapinella (Paxillus) panuoides

Cap (1–4 in. wide) oyster-like on coniferous wood, spring to fall in NA, EU. Indistinct odor and taste. Edibility? The genus *Tapinella* has just two species in Cascadia, *T. atrotomentosa* (see key lead 307a [page 246]) and *T. panuoides*.

↓328b. On wood; large, velvety stipe (see key lead 307a [page 246])

↓328c. On wood; stipe ± absent; cap ± white; gills do not rub off

Crepidotus applanatus var. *applanatus*

Many *Crepidotus* species fruit on hardwoods or conifers in spring–fall (NA, EU). The odor is generally mild. None are edible.

↓328d. On wood; stipe ± absent; cap ± ochre with fibrillose scales

Crepidotus calolepis (see also key lead 1a #3 [page 42])

Cap (< 3 in. wide) has a gelatinous texture. Usually on hardwoods year-round in NA, EU. Mild odor and ± mild taste. *C. mollis* (EU) is similar. Not edible. *C. crocophyllus* (NA, EU) is also ochre (to brilliant yellow or orange) and covered with dark brown scales, but does not have a gelatinous layer in the cap.

328e. Not on wood *329a*

329a. (328e) Stipe cartilaginous with a long tapered "root" *330a*

↓329b. Stipe not cartilaginous, not rooting; gills ± decurrent, bright yellow

Phylloporus arenicola (a gilled bolete)

Cap (1–2 in. wide) suede-like. Under two-needle pines and Douglas fir, spring to fall. Two unnamed species are also in Cascadia, but not *P. rhodoxanthus* (EU). Mild odor and taste. Edibility? Gills rub off easily.

↓329c. Gills decurrent, bruise reddish; cap (2–12 in. wide), margin inrolled

Paxillus cf. *obscurisporus*

Four EU species are known from Cascadia: *P.* cf. *obscurisporus* (spore print red-brown), *P. involutus* (spore print reddish ochraceous, page 321), *P. cuprinus* (with Betulaceae, page 321), and *P.* aff. *ammoniavirescens* (strong green with ammonia-like odor). In all four, cap margin long inrolled; gills rub off; odor pleasant; and taste mild to unpleasant. Seriously poisonous if eaten repeatedly.

329d. Cap (3–12 in. wide) golden; cap and stipe with granulose coating

Phaeolepiota aurea

A unique and distinctive species growing in clusters on the ground, typically on roadsides near either conifers or hardwoods in NA, EU, Asia. Odor strong and pungent, unpleasant; taste mild to astringent. Not a good edible.

330a. (329a) Olive-green to olive-brown cap with a glutinous surface

Phaeocollybia olivacea group

Cap < 4 in. wide. Odor when fresh ± of radish or cucumber, taste mild to bitter. Viscid and cartilaginous. Edibility? *Phaeocollybia* species are common in the coastal forests of Cascadia, where at least twenty-four species can be found, some reaching diameters of 8 in. Five or six species resemble *P. olivacea*. None are edible.

↓330b. Brownish orange hygrophanous cap (< 2 in. wide), low umbo

Phaeocollybia attenuata group

Young gills pinkish buff to lilac, age dark orange-brown. Stipe slender, ≤ 3 in. above ground, 9 in. below ground. Many localities in Cascadia, usually under Sitka spruce or redwoods. Odor complex and variable, often of radish, sometimes floral. Taste disagreeable. Not edible.

330c. Large size; caps (2–8 in. diameter) ± glutinous; robust stipe

Phaeocollybia spadicea group   *Phaeocollybia kauffmanii*

A dozen drab, ochraceous to brown species fit this general description. The thick, firmly stuffed cartilaginous stipes are never hollow. Microscopy is needed to differentiate the species. They fruit in fall almost always in closed canopy, coniferous rainforests of Cascadia, typically with a high humus content. Odors vary from mildly to strongly farinaceous. Taste bitter to farinaceous. Not edible.

331a. (327g) Cap (< 2 in. wide) ± striate when moist, ± viscid, ± fibrous or membranous veil; hygrophanous; on wood, moss, or humus (*Galerina* spp.) *333a*

↓331b. Cap (< 1 in. wide) dry, ± striate, whitish to cinnamon-colored; in grass, on soil, on dung; if on wood, ring present; gills free or attached, cinnamon-colored (*Conocybe, Pholiotina* spp.) *332a*

331c. Cap (< 2½ in. wide) slippery, striate, often dissolving; gills ± free

Bolbitius cf. *callistus*   *Bolbitius titubans* var. *titubans*

B. titubans var. *titubans* (= *B. vitellinus*) appears spring to fall in farmyards, gardens, grass in NA, EU, Asia, North Africa. *B. callistus* (NA) is viscid, often reticulate, blue when very young but quickly olive-yellow to red. Both are nonpoisonous with an indistinct odor and taste.

332a. (331b) Cap (< 1 in. wide) whitish to pale cinnamon color; fragile

Conocybe apala (± life-size)

Cap ± wrinkled, flesh and stipe all very thin. Common in grass, gardens, and meadows in spring to fall in NA, EU, Asia, North Africa. Spore print rusty brown. Indistinct odor and taste. Not toxic.

↓332b. Cap (< 1 in. wide) orange-brown to cinnamon-colored; stipe color same as cap; not very fragile

Conocybe 'tenera' group (± half life-size)

In grass or in woods, spring to summer, sometimes fall, in NA, EU. Indistinct odor and taste. Edibility? Several unrelated look-alikes all pass for *C. tenera*. Looks like deadly *Pholiotina rugosa*, which has a partial veil and often a ring on the stipe.

332c. Cap (< ½ in. wide) moist, orange-brown; membranous veil, ± a ring

Pholiotina cf. rugosa

Long known as *P. filaris/Conocybe filaris* (EU), with ± six species in Cascadia. The gills and spore print are rusty brown. If the ring disappears, resembles *C. tenera*. Found on rich clay soils, on wood chips, or in grass on buried wood, summer to fall. All are potentially deadly (amatoxins). Similar deadly species may be found in both the Northern and Southern Hemispheres.

Note: *Galerina* is a challenging genus containing more than three hundred species worldwide and at least five subgenera that may be raised to species rank. All are small to tiny, tan to brown mushrooms. Three Cascadia region species in the subgenus *Naucoriopsis* contain Amanitins. One, *Galerina marginata*, has caused one deadly poisoning and contains concentrations of alpha- and beta-Amanitins comparable to the levels found in the deadly *Amanita phalloides*. *G. castaneipes* contains slightly lower concentrations of alpha-Amanitins and usually also contains beta-Amanitins. *G. badipes* lacks alpha- and beta-Amanitins but reportedly contains gamma-Amanitins. Positive identification of the deadly species is exceptionally challenging. No *Galerina* species should ever be consumed, even though no amatoxins have been identified in *Galerina* subgenera other than *Naucoriopsis*.

333a. (331a) Grows in moss (page 262) *335a*

333b. Sometimes in moss, typically on wood or high-lignin soils *334a*

334a. (333b) Cap (< 2½ in. wide), hygrophanous, ± viscid, ± ring; ± membranous veil

Galerina marginata

Cap skin ± peelable. Otherwise much like *G. autumnalis* (EU). Found on coniferous wood, summer to fall, sometimes on hardwoods, sometimes on moss in NA, EU, Asia. Odor ± strongly farinaceous. Do not taste, contains alpha-Amanitin and is DEADLY. *G. venenata* is a synonym. Member of subgenus *Naucoriopsis*.

↓334b. Cap (< 2 in. wide) ± viscid, ± no ring; ± thin fibrillose veil

Galerina castaneipes

Faint fibrillose veil, generally no ring. Odor and taste indistinct. Known from old logs of oak and madrone as well as on moss in Cascadia. *G. badipes* (distinguished microscopically) also has a fibrillose veil. Both are in subgenus *Naucoriopsis*, and both may be DEADLY.

334c. Cap (< 1 in. wide) viscid, hygrophanous, amber-brown; no veil or ring

Galerina near *sideroides*

The image is of an unnamed species with DNA very close to (near) *G. sideroides* but with a non-viscid, non-hygrophanous cap. On rotten oak logs, fall. (*G. sideroides* and the similar *G. mammillata* are on conifers in NA, EU.) In subgenus *Sideroides*. The odor and taste are mild. Do not eat.

335a. (333a) Cap (< 1 in. wide); white fibrillose veil, ring zone on stipe; in sphagnum bogs

Galerina paludosa

The white fibrillose veil distinguishes this from other *Galerina* species in sphagnum bogs. Fruits spring to fall in NA, EU, Asia. Odor and taste indistinct. Do not eat. In section *Mycenopsis* in subgenus *Galerina* of the genus *Galerina*.

↓335b. Cap (< 1 in. wide) moist, broadly convex to flat; stipe darker at base, downy

Galerina vittiformis group

No veil; flesh thin and fragile. Among mosses and on mossy logs, late summer to fall in NA, EU, Asia. Indistinct odor and mild taste. Too small to consider eating, and this is a dangerous, large genus where microscopy is needed to confirm identifications. There are at least three distinct species, all in subgenus *Galerina* of the genus *Galerina*.

↓335c. Cap (< 1 in. wide) hygrophanous; stipe with remnants of fibrillose veil

Galerina semilanceata group

Robust for such a small species. Stipe pinkish buff with a whitish fibrillose coating below where the veil had been. Veil evanescent. In moss, grass with moss, and on mossy logs, spring to fall in Cascadia. Odor and taste mild. Do not test it for edibility. In subgenus *Tubariopsis*.

335d. Cap (< ⅓ in. wide) hygrophanous; thin fibrillose veil but no ring zone

Galerina cf. *subfiliformis*

Sharply conic cap, not expanding much in age (NA). Stipe pruinose at top; sometimes a few veil remnants on lower stipe. Positive ID requires microscopy (as with all these NA moss-inhabiting species). No data on season, odor, or taste. In section *Mycenopsis* of subgenus *Galerina*. Not edible.

336a. (327f) Species not colored or staining blue-green (page 265) *338a*

336b. Species colored or staining blue-green (page 264) *337a*

337a. (336b) Cap (2–5 in. wide) greenish to orangish, bluish green bloom; no veil

Gymnopilus punctifolius

Cap is dry; sometimes has lilac-colored mycelium at stipe base. Distinctive bluish green bloom is hard to capture on film or digitally. Distinguished by absence of any veil. In coniferous woods, summer to winter in NA. Odor indistinct, taste very bitter. Not known to be hallucinogenic.

337b. Cap (2–8 in. wide) yellow, salmon-pink, wine-red, ± dense dark reddish scales; ± persistent veil

Gymnopilus luteofolius (= G. aeruginosus; above and at right)

Two ± different-looking fungi have the same DNA, so have the same name. The cap and stipe may both be densely scaly. Blue staining in this case indicates the presence of psilocybin. On both hardwood and conifer logs in NA. Odor mild to alkaline. Taste bitter.

Gymnopilus ventricosus (see description for key lead 338a)

338a. (336a) Cap (2–10 in. wide) ± orange-brown; cespitose; near trees; persistent veil/ring

Gymnopilus voitkii

Two species, *G. ventricosus* (western NA, page 264) and *G. voitkii* (western and eastern NA) are distinguished by spore size. Reports of psilocybin in Cascadia species are unconfirmed. They do not stain blue, are bitter, and are found near or on living trees or conifer wood in the fall. In eastern NA, big laughing gym (*G. junonius*) is a blue-staining, hallucinogenic species.

↓338b. Cap (< 2 in. wide) bald, dry; slight cobwebby veil; stipe stains brown

Gymnopilus cf. *flavidellus*

A member of the smooth-capped *G. penetrans* (NA, EU) group. *G. flavidellus* has a bald cap with a smooth margin. Gills initially pale yellow. Yellow stipe stains fulvous brown. On coniferous and deciduous wood, fall to winter in NA. Odor ± radish-like, taste bitter. Inedible.

↓338c. Cap (< 2 in. wide) bald, dry; slight veil; stipe remains yellow (with white rootlets)

Gymnopilus cf. *aurantiophyllus*

G. aurantiophyllus is in the smooth-capped *G. penetrans* (NA, EU) group. The cap often has a wavy margin; gills initially pale yellow; stipe yellow. On coniferous and deciduous wood, summer to fall in Cascadia. Indistinct odor, very bitter taste. Inedible.

338d. Cap (< 1½ in. wide) moist, finely scaly to bald; no veil

Gymnopilus oregonensis

Cap convex to bell-shaped, hygrophanous, cinnamon-brown to yellow-brown, margin often lighter. Gills yellow, darkening from spores and may stain rusty. Stipe white-frosted, yellow-brown to red-brown underneath. Mainly on conifer wood, summer to fall in Cascadia. Indistinct odor, taste mild to slightly bitter. Edibility? Long known as *G. picreus*, a (EU) sister species.

339a. (327e) Not growing on burned wood or ground *341a*

339b. Growing on burned wood or ground *340a*

340a. (339b) Cap (< 1½ in. wide) hygrophanous, moist, reddish brown to tawny; veil whitish

Crassisporium funariophilum

The pallid veil leaves white remnants on the cap margin. Cap bald, feels soapy, stipe cinnamon-colored. Summer in western NA. Indistinct odor and taste. *Pholiota molesta* (Cascadia) is a viscid burn species with a similar cap color and a fatter, whitish stipe; whitish veil leaves cinnamon-colored patches on stipe.

↓340b. Cap (1–3 in. wide) hygrophanous, viscid, yellow-brown to reddish

Pholiota brunnescens

Cap with low umbo, and margin with scattered white veil remnants, lemon-yellow, on stipe. Gills attached, whitish, yellow-brown in age with pallid edge. Scattered to dense on burned soil, wood, spring to fall in NA, China. Somewhat disagreeable odor and taste. Edibility?

340c. Cap (< 2 in. wide) hygrophanous, viscid, yellow-brown to cinnamon-colored

Pholiota carbonaria

Cap without an umbo, edge ± scattered pallid veil remnants, appressed fibrillose. Gills pale brown, aging brown. Year-round for up to two years post fire in NA, EU, Asia, North Africa. Slightly musty odor and taste. Known to be poisonous. Synonymous with *P. highlandensis* and *P. fulvozonata*.

341a. (339a) Cap not hygrophanous *343a*

341b. Cap hygrophanous; usually growing in clusters (*Kuehneromyces* spp.) *342a*

342a. (341b) Cap (< 1½ in. wide) pale to dark butterscotch, ± viscid; spring

Kuehneromyces lignicola

Bald cap has a small umbo and veil remnants on the margin when young. Gills pale, dingy tan, aging cinnamon-buff. Stipe butterscotch color, stained dull, rusty brown from base upwards. Stipe may have veil remnants, but prominent scales are absent. Odor and taste mild. Resembles deadly poisonous *Galerina* species. Do not eat. Range NA, EU, Asia.

342b. Cap smooth, red-brown to yellow-brown; scaly stipe below a ring

Kuehneromyces mutabilis

Cap broadly umbonate, moist to viscid, veil remnants on margin. Partial veil whitish, membranous to fibrillose, leaving distinct to indistinct ring. On hardwoods and conifers or buried wood, year-round, mainly fall, in NA, EU, Asia. Odor and taste indistinct to pleasant. Edible, but avoid because deadly *Galerina* species look similar.

343a. (341a) Cap and/or stipe viscid; cap and stipe bald, silky, or scaly (page 269) *346a*

343b. Neither cap nor stipe obviously viscid, at least when young *344a*

344a. (343b) Obvious on wood or on buried wood (page 268) *345a*

344b. Appears terrestrial; cap (1–4 in. wide) and stipe scaly, dingy brown

Pholiota terrestris

Cap and stipe color vary from light to dark brown, margin typically with hanging veil remnants. Along roads and disturbed areas, in lawns, rarely on wood, summer to winter in NA, Japan. Odor and taste mild. Edible but flavorless.

345a. (344a) Cap (1–4 in. wide) and stipe yellow to orange

Pholiota flammans

A very distinctive species that is sometimes a bit more brownish. Cap scales may wear off when old. Gill edges may brown when rubbed. Grows on rotten conifer wood, especially hemlocks, summer and fall in NA, EU, Asia. Odor mild to fruity or unpleasant, taste mild to peppery. Edible but poor quality.

↓345b. Cap 1–4 in. wide; young gills whitish, aging brownish without green tint

Pholiota squarrosoides (above and at right)

Cap viscid beneath the scales. Gills whitish, slowly rusty brown. Stipe stains rusty brown near base. Found on old alder and maple logs, sometimes on conifer wood, late summer to fall in NA. Odor mild to sweet and taste indistinct. Edible and bland, but rub off scales before eating. Can cause mild GI distress. A very similar species, long thought to be common in Cascadia, *P. squarrosa*, has slightly larger spores and sometimes green-tinted gills, but has not yet been verified here. *P. squarrosoides* and *P. squarrosa* have markedly different DNA.

345c. Cap 1–3 in. wide, with irregularly contorted gills; spores not discharged

Pholiota nubigena

The cap is slightly viscid, yellow-brown to white. Fruits near melting snowbanks in subalpine zone, persists until fall in Cascadia. On conifer wood. Odor sweet, ± wintergreen-like, taste mild. Edibility?

346a. (343a) Cap scaly, at least when young (page 270) *349a*

↓346b. Cap bald; veil leaves a ring on stipe (page 270) *348a*

346c. Cap bald; no ring (may have veil remnants on margin) *347a*

347a. (346c) Cap (< 2 in. wide) bright pinkish orange; stipe pale yellow; no scales

Pyrrhulomyces astragalinus

Formerly *Pholiota astragalina*. Distinctive. Cap margin with veil remnants when young. Gills yellow, staining yellow-brown. Veil pallid yellow, barely detectable ring zone. On rotting conifers, late summer to fall in NA, EU, China. Odor not distinctive, taste bitter. Inedible.

↓347b. Cap (2–6 in. wide) yellowish to rusty; ± veil remnants on margin

Flammula cf. *malicola* var. *macropoda*

The specimen shown in the image is darker than the normal dingy, ochraceous-tawny mushroom. The stipe widens at base and stains dark rusty brown. On or near hardwood or conifer trees in NA. Odor of green corn, taste mild. Not edible. In the *F. alnicola* group of smooth-capped, difficult-to-ID species.

347c. Cap (1–3 in. wide) color variable, tawny ochre center, greenish yellow margin

Pholiota spumosa group

Partial veil is pale yellow and delicate. Young gills yellow to greenish, aging rusty brown. Usually on conifers and wood chips, fall to winter in NA, EU, Asia, North Africa. Odor and taste mild, sometimes green corn–like. Flesh is thin and soft. Edibility?

348a. (346b) Cap (1–3 in. wide) pale to dark red-brown; gills white at first

Stropharia albivelata

Formerly in genus *Pholiota*. Cap has a viscid pellicle that may wrinkle. Upper part of ring is striate. Scattered on debris under conifers, fall in Cascadia. Odor fungus-like and taste mild. Edibility?

348b. Cap (1–3 in. wide) ± viscid, ± tawny; veil remnants on margin and stipe

Flammula alnicola

Formerly in genus *Pholiota*. Cap convex to flat, often with faint yellow veil remnants on margin. Ring on stipe may disappear. Stipe fibrillose below ring zone. In clusters on and near hardwoods and conifers, summer to fall in NA, EU. Odor fruity, taste mild. Not edible. Highly variable.

349a. (346a) Generally found only on hardwoods *350a*

↓349b. On hardwoods or conifers; cap 2–6 in. wide, tawny, yellow-orange, or orange; large, removable scales

Pholiota adiposa (formerly Pholiota aurivella) group

Pholiota aurivella

Pholiota limonella

Recent DNA evidence has shown that *P. adiposa* is highly variable and now includes what were once known as *P. aurivella* and *P. limonella*. A white partial veil leaves a fleeting ring, pale gills age rusty brown. Summer to fall in EU, NA. Odor sweet, taste slight. Edible.

↓349c. Cap 2–6 in. wide, viscid, yellow to orange, dark scales; scaly stipe

Pholiota adiposa (cultivated)

The Columbia Mushroom Company gifted me a grow kit for *P. adiposa* with a chestnut/cinnamon-colored cap. It is now one of my top two favorite cultivated edibles. Even the stipes are delicious. As with all edible species of fungi, a few people suffer GI distress after eating.

349d. Usually on conifers; cap dark center, pale margin, small scales

Pholiota decorata

Cap (1–3 in. wide) fibrillose streaked. Gills whitish, aging dingy clay color. Stipe pallid, top silky, fibrillose below ring zone. Usually on conifer branches, rarely hardwood, in western NA. Odor faintly fragrant, taste mild. Not edible.

350a. (349a) Cap (2–6 in. wide) reddish brown to brown; white-edged gills

Hemistropharia albocrenulata

Once placed in genus *Pholiota*. The cap has scattered, brown, fibrillose scales. Brownish partial veil is fibrillose-cottony. The stipe is grayish above, dark brown below, with a fleeting ring zone. Usually on maples alive or dead in NA, China. Odor indistinct and taste mild to bitter. Harmless.

350b. Cap (3–8 in. wide) white to creamy buff; whitish to buff scales

Hemipholiota populnea (= *Pholiota populnea*)

Formerly *P. destruens*. Whitish veil remnants on the cap margin. The stipe is thick and hard. Found on poplar, cottonwood, and willow, late summer to fall in NA, EU. Odor fungoid, taste ± bitter. Edible but poor flavor and very tough.

351a. (327d) Cap 1–4 in. wide; stipe > ⅓ in. tall; yellow-brown to brown; no veil

Agrocybe smithii/A. putaminum

Distinguished from the other medium-sized *Agrocybe* species by the absence of a veil. Found spring to summer in wood chips and mulch beds in NA. The odor and taste are ± farinaceous. Edibility? *A. putaminum* appears to be in Cascadia, EU. *A. smithii* (NA) may or may not be a separate species.

↓351b. Cap (1–4 in. wide) cream to brownish, smooth to cracked; membranous veil

Agrocybe praecox group

The cap is very soft and smooth. The veil is thin but distinctly membranous and usually leaves a ring. In fields, woods, chip piles, or mulch, spring to fall in NA, EU. Odor mild to farinaceous, taste often ± bitter. Edible, not great. Three or four very similar species in Cascadia.

↓351c. Cap (1–4 in. wide) whitish to brownish, smooth to cracked; fleeting ring zone

Agrocybe dura/A. molesta

Appearance strongly overlaps with *A. smithii* and *A. praecox*. Distinguished by being a grass decomposer (in NA, EU, Asia, North Africa). Odor indistinct, taste mild to bitter. Edible but often with unpleasant after taste.

351d. Cap (< 1½ in. wide) cream to yellowish cream; fibrillose veil but no ring

Agrocybe pediades/A. semiorbicularis

Mycologists debate whether two-spored *A. semiorbicularis* should be considered distinct from four-spored *A. pediades*. They grow in grass, on manure, or on bare ground, spring to fall in NA, EU. They have the same ITS DNA. Odor and taste ± farinaceous. Not edible.

352a. (327c) Cap (1–3 in. wide) viscid, whitish to ochre; white gill edges

Hebeloma velutipes

Many confusingly similar species are in Cascadia, some are toxic. *H. crustuliniforme* (NA, EU) is one of many *Hebeloma* species in Cascadia. With hardwoods and conifers, low elevations, fall in NA, EU. Odor ± of radish, taste astringent to bitter.

↓352b. Cap (< 2½ in. wide) two-toned white and brown, ± viscid; sweet smell

Hebeloma sacchariolens group

Under conifers and hardwoods and on waste ground at low elevations, fall in NA, EU, New Zealand. The burnt-sugar or orange-blossom odor is distinctive. Taste mild, sweet, or bitter. Sometimes eaten, and no reported poisonings so far. Nevertheless, not recommended.

↓352c. Cap (< 2½ in. wide) two-toned grayish brownish; white fibrillose veil

Hebeloma mesophaeum group

Cap bell-shaped, ± umbonate in age, often with patches of pallid fibrillose veil material midway to the disk. Gill edges white flocculose. Under conifers, often in grass near trees, spring to fall in NA, EU. Odor and taste mild to radish-like. Poisonous. Possibly eight species with very similar DNA are in Cascadia.

352d. Cap (1–3 in. wide) fawn-colored to red-brown, viscid; in sphagnum bogs

Hebeloma incarnatulum

Growth near conifers in sphagnum bogs is distinctive. No partial veil. The stipe is always long and slender. Odor and taste are of radish. Edibility? A member of the *H. velutipes* group. Range NA, EU.

Note: The formerly huge genus *Inocybe* has been split into four genera. Members of these genera are all ± thready looking, with ± dry caps. About ninety named species are in Cascadia alone. These species should be treated as seriously poisonous, as most contain significant amounts of muscarine.

353a. (327b) Cap smooth to matted fibrillose, rarely splitting radially (page 276) *358a*

353b. Cap fibrillose to scaly, frequently splitting radially *354a*

354a. (353b) Odor ± mild, not of green corn, raw fish, or pine resin *357a*

↓354b. Odor ± spermatic or of raw fish and pine resin *355a*

354c. Odor of fresh green corn; cap (1–4 in. wide) sharply conical, straw color

Pseudosperma sororium group

No veil. In mixed woods, often with Douglas fir, summer to fall in NA. The ten look-alikes in this group do not include *P. rimosum* (= *Inocybe rimosa*), which has a spermatic odor and has not been found in Cascadia.

355a. (354b) Base of stipe not blue-green *356a*

355b. Base of stipe blue-green; cap (≤ 1½ in. wide) scaly, rarely splitting

Inosperma calamistratum group

Inosperma maximum

Inosperma calamistratum was formerly known as *Inocybe calamistra* (NA, EU, Asia, India), and *Inosperma maximum* was formerly *Inocybe hirsuta* var. *maxima*. *Inosperma maximum*, distinguished by a much fatter stipe than *I. calamistratum* plus gills and flesh that redden when bruised, is often found under hemlocks in Cascadia. Ten named and unnamed *Inosperma* species in Cascadia, with fishy, resinous, geranium-like, or green corn–like odor (*I. mucidiolens*, NA). Edibility? All considered poisonous (no psilocybin).

356a. (355a) Cap (< 2 in. wide) with prominent umbo, gray-brown; whitish cortina

Inocybe fuscidula

Gills light beige with a white edge. Upper half of ± fragile stipe pruinose, lower ± fibrillose. Under hardwoods and conifers, summer to fall in Cascadia, EU. Odor spermatic to bleach-like, taste indistinct to unpleasant. Poisonous.

356b. Cap (< 1½ in. wide) with umbo, milky-coffee to gray-brown; whitish cortina

Inocybe subdestricta sensu Stuntz

Gills are light beige with a white edge. Stipe fragile, upper half pruinose, base ± bulbous. Under hardwoods and conifers, summer to fall in NA. Odor spermatic to bleach-like, taste mild to unpleasant. Poisonous.

357a. (354a) Subalpine; cap (< 2 in. wide) and stipe ochre; white brief cortina

Mallocybe fibrillosum (= M. subdecurrens)

Long known as *Inocybe dulcamara*. The cap is flat, not umbonate, and both cap and stipe are rough-scaly. The gills are initially light yellow-brown and fringed white or brown. Odor is faintly radish-like or fishy. Taste slightly sweet to slightly bitter. Poisonous.

↓357b. Cap (< 1½ in. wide) with slight umbo, ochre to brown; brief tan cortina

Inocybe lacera

Fruits in spring often on sandy or burnt ground under aspens and conifers in NA, EU. The odor is faintly spermatic or bleach-like or faintly like green corn. Taste indistinct but avoid tasting due to high muscarine levels. Poisonous.

↓357c. On rotten wood; cap (< 1¼ in. wide) finely scaly, dark brown; cortina pallid

Inocybe lanuginosa

Cap broadly umbonate, darkest in center with upcurved, pointed scales; top of stipe finely sanded, lower stipe like cap. Fleeting veil. Associated with conifers, spring to winter in NA, EU. Indistinct odor and taste. Poisonous.

357d. Cap (< 3 in. wide) finely scaly, yellow-ochre to brown; cortina tan, fleeting veil

Inocybe olympiana

West of the Cascade Mountains in old growth forests, late summer to fall. The base of the stipe has a slightly marginate bulb. Gills whitish when young, retaining whitish margin in age. Odor and taste indistinct to mildly farinaceous. Poisonous.

358a. (353a) Odor distinctive, often but not always spermatic *359a*

358b. Odor indistinct; cap (< 2 in. wide) ± umbonate, yellow-brown; no veil

Inocybe praecox

Limited to western WA in low-elevation, mixed forests, spring to summer. Stipe may have a marginate basal bulb. Gills pallid when young, edges remain pallid. Indistinct odor and taste. Presumed poisonous. In *I. splendens* complex. *I. castanea* (NA) is similar in shape, color, habitat, and toxicity.

359a. (358a) Strong, fruity to rancid odor; cap (< 2 in. wide) brown; turnip-bulb stipe

Inocybe napipes/Inocybe mixtilis group

Cap silky-fibrillose, ± conic to umbonate. Gills whitish to pale brown (edges remain white in *I. napipes*). Bulbous stipe base (marginate in *I. mixtilis* group). Cortina present in *I. napipes*. Found in many forest types, usually fall in NA, EU. All are poisonous.

↓359b. Distinct spermatic or bleach-like odor; cap (± < 1½ in. wide) ± white, umbonate; white cortina; whitish, silky stipe

Inocybe geophylla complex

Inocybe pallidicremea (= *I. lilacina*)

Inocybe whitei/I. pudica

Inocybe posterula

I. geophylla complex includes five of the twenty-nine NA, EU species in Cascadia with conspicuous, but disappearing, white cortina and no stipe bulb. Under conifers and hardwoods, spring to winter. Roughly five species resemble *I. pallidicremea* (NA) and have ± lilac caps. *I. whitei* (= *I. pudica*) (NA, EU, "correct" name is in debate) is larger (< 3 in. wide) and blushes pink, red, or orange, but never lilac. It is probably the most common *Inocybe* species in Cascadia. *I. posterula* (NA, EU) is best distinguished microscopically from *I. geophylla* (NA, EU), though its odor is only faintly spermatic to bleach-like.

359c. Wet-dog odor; cap (< 2½ in. wide) ± greasy, white to pale brown

Inocybe sindonia group

Disappearing veil often leaves hanging fragments on cap margin but not on cap itself. Under conifers or hardwoods, fall in NA, EU. Gills white to cream-colored when young. Odor and taste farinaceous to bleach-like. Inedible.

Note: The genus *Cortinarius* is now divided into seven genera present in Cascadia: *Cortinarius, Aureonarius, Calonarius, Cystinarius, Hygronarius, Phlegmacium,* and *Thaxterogaster.* Subgenera *Telamonia* and *Dermocybe* remain in genus *Cortinarius.* The other former subgenera are split into multiple new groupings. The entries that follow provide key leads to a few of the species found in Cascadia. Nearly all have a cortinate (cobwebby) veil when young, which leaves fine, thread-like fibers on the stipe that catch the distinctive orangish-brown spores. Many distinctive odors, gill colors, flesh colors, and color changes with KOH (lye, a caustic, use caution). Many glow in the dark in interesting colors under UV light.

360a. (327a) Associated with oaks (page 289) *380a*

360b. Associated with conifers or hardwoods other than oaks *361a*

361a. (360b) Caps small, dry, and fibrillose; gills brightly colored; ± UV active (subgenus *Dermocybe*) (page 287) *377a*

↓361b. Viscid caps; ± dry, often very thick stipes, with ± enlarged stipes base (genera *Cortinarius, Calonarius, Phlegmacium*) (page 286) *375a*

↓361c. Both cap and stipe viscid (subgenus *Myxacium*) (page 284) *372a*

↓361d. Silky-satiny, non-hygrophanous species (genera *Cortinarius* and *Phlegmacium* and new subgenera/sections) (page 283) *370a*

↓361e. Smooth, dry caps, sometimes hygrophanous, often blackening in KOH; ± UV active (subgenus *Telamonia*) (page 280) *364a*

↓361f. Caps dry, non-hygrophanous, usually scaly or hairy, glow yellow under UV light (genera in Cascadia: *Cortinarius, Aureonarius, Cystinarius,* and two new subgenera) (page 280) *363a*

↓361g. With contorted gills that do not forcibly discharge spores (page 280) *362a*

↓361h. Cap (< 6 in. wide) and stipe rough and scaly; all parts purple-black

Cortinarius violaceus

The only known member of subgenus *Cortinarius* genus *Cortinarius* in Cascadia. This is a distinctive species, like nothing else (EU, NA). It is found under both hardwoods and conifers at all elevations, late summer to fall. The odor is weakly like cedar and the taste may be slightly peppery. Edible, though not desirable. Be sure of identification before eating.

361i. Cap (2–6 in. wide) ± tan, frosted when young; membranous veil

Cortinarius caperatus

The only *Cortinarius* species with a membranous partial veil, usually leaving a ring on the stipe, subgenus *Rozites* (NA, EU, Africa). In coniferous woods, late summer through fall. The odor and taste are usually pleasant. Edible (discard the tough stipes).

362a. (361g) Cap (< 2 in. wide) greasy, smooth, dingy brown; gills distorted

Cortinarius (Thaxterogaster) pinguis

Found in the mountains often near spruce, late summer until snowfall in western NA. Contorted gills yellowish, aging rusty brown. (For white contorted gills, see *Russula chlorineolens* [page 321] and *R. suboculata* [page 321].) Yeasty odor, ± unpleasant taste. Not edible.

↓362b. Cap (2–4 in. wide) pale yellow to brown; dense yellowish cortina

Calonarius saxamontanus
(= *Cortinarius saxamontanus*)

The contorted gills are covered by a persistent tough, yellowish veil. The thick stipe is yellowish with a bulbous base. Under pine and true fir above 4,500 ft., summer in western NA. Odor fungoid. Not edible.

362c. Cap (1½–3 in. wide) and stipe white; dense, persistent white veil

Calonarius magnivelatus
(= *Cortinarius magnivelatus*)

Rare north of the Sierra Nevada mountains. Fruits after snowmelt above 4,500 ft., buried under true firs and pines in humps in duff in western NA mountains. Veil never opens. Indistinct odor and taste. Edibility? *C. wiebeae*, brown gills when young. *C. magnivelatus*, whitish gills when young.

363a. (361f) Cap (< 2½ in. wide) yellow-green to golden brown, small central scales

Cortinarius clandestinus group

Brown to black scales. Yellow fluorescence. At least four conifer species plus an oak associate (see key lead 383e [page 291]), were all once called *C. cotoneus*. Common. Remains of a universal veil on stipe. All elevations, spring-fall in NA. Odor ± of radish. Taste and edibility not recorded.

↓363b. Cap (1–3 in. wide) ± rust-brown, dark reddish brown scales; in deep moss

Aureonarius rubellus (formerly *Cortinarius rubellus* = *C. rainierensis*, *C. speciosissimus*, *C. orellanoides*)

Gills yellow-brown when young. Stipe with yellowish veil bands. Sometimes on rotten, mossy logs, summer to fall in NA, EU. Rare. Blue UV fluorescence. Odor and taste ± of radish. Contains orellanine: DEADLY. Slow-acting toxin destroys the kidneys. If eaten, may require kidney transplant.

363c. Cap (1–3 in. wide) yellow-brown and brown; yellow, membranous universal veil

Cortinarius parkeri

Small volva at base of pallid stipe. Cap a bit hygrophanous with appressed fibrillose scales. Bright yellow under UV light. Conifers and mixed forests, all elevations, spring to summer in Cascadia. Odor strong, ± of radish; taste mildly of radish. Edibility?

364a. (361e) Fruits in late summer to fall (page 282) *366a*

364b. Fruits near snowbanks or two or three weeks after snowmelt *365a*

365a. (364b) Cap (1–3 in. wide) ± brown; a greenish white to yellow-buff universal veil; strong yellow UV fluorescence

Cortinarius colymbadinus group

Cortinarius ahsii

Cortinarius flavobasilis

Cortinarius bridgei

Cortinarius colymbadinus

All four species are found in the mountains of western NA. They are hygrophanous with flesh color like the surface color or sometimes with red vinaceous tints. *C. colymbadinus* (also in EU) shows the yellow fluorescence in all parts, while for *C. ahsii* only the universal veil is yellow. For *C. bridgei* and *C. flavobasilis* (distinguished microscopically), the stipe base fluoresces orange and is instantly red in strong KOH. All have a reddish reaction of universal veil to 10 percent KOH.

365b. Cap (1–3 in. wide) reddish brown to dark brown; no distinct UV fluorescence

Cortinarius politus

Watery gray to gray-violet streaks in the upper stipe. Lacks a universal veil. Like *C. colymbadinus* group, under conifers at mid- to high elevations, known from WA. Odor fungoid. *C. brunneovernus* (WA, CA), another brown spring species, lacks violet tints in stipe.

366a. (364a) Associated with conifers or in mixed woods *367a*

366b. Cap (< 4 in. wide) vinaceous gray; under cottonwoods and aspens

Cortinarius lucorum

Hygrophanous. Young gills purplish or magenta. Stipe cylindric to bulbous. Thick universal veil sheathes lower stipe. In mixed lowland forests, summer to fall, northern NA, Norway. Odor mild to fungoid; taste fungoid to slightly bitter. Edibility?

367a. (366a) Cap shades of brown to black without orange or red tones *369a*

367b. Cap orange-brown to reddish *368a*

368a. (337b) Cap (1–3 in. wide) hygrophanous, reddish brown; gills ± orangish

Cortinarius californicus group

Distinguished from brightly colored *Dermocybe* species by a hygrophanous cap that fades to brown. Orange cortina. Mixed forests, fall in Cascadia. Stipe ± cap color. Odor and taste mild to radish. Edibility?

↓368b. Cap (< 1½ in. wide) ± hygrophanous, ± orange-brown; gills yellow

Cortinarius gentilis

Believed deadly (it isn't), but is very similar to deadly *C. rubellus*. Gills age brownish orange. Stipe with yellow bands from universal veil. Common in late summer to fall under conifers in NA, EU. Indistinct to radish-like odor and taste.

368c. Cap (1–2½ in. wide) glossy, vivid orange-brown; gills pallid; no veils

Hygronarius 'renidens' (= *Cortinarius 'renidens', Gymnopilus terrestris*)

The Cascadia species is distinct from (EU) *H. renidens.* Young gills pallid, edges may remain white. Lacks a cortina and thus was once placed in genus *Gymnopilus.* Under conifers in the mountains. Mild odor and taste. Edibility?

369a. (367a) Cap (1–4 in. wide) streaked dark to pale cocoa-brown; fleeting veil

Cortinarius cacaocolor

The large *C. brunneus* group contains several similar members. Young gills are light brown, aging rusty brown. In coniferous forests, fall in Cascadia. Odor like rhubarb; taste mild. Edibility? *C. brunneus* (NA, EU) is similar, but cap is smoother.

369b. Cap (1–2½ in. wide) dark brown with yellow tints; sparse yellow universal veil

Cortinarius nauseosouraceus

Hygrophanous. Gills yellow-brown when young, retaining yellow edge. Low-elevation coniferous forests, fall in Cascadia. Odor strong, varying from earthy to green corn–like. Not tasted. *C. uraceus* (NA, EU) has greenish tints and lacks the strong odor. Edibility?

370a. (361d) Cap and/or stipe with blue to violet colors (page 284) *371a*

370b. Cap (< 4 in. wide) and stipe silvery whitish aging pale brown; gills pale brown

Cortinarius alboglobosus

White universal veil leaves whitish bands on stipe. Flesh very pale brown. Under pine, spruce, and fir, late summer to fall in NA, EU. Odor earthy, taste slightly unpleasant. Edibility?

371a. (370a) Cap (1–4 in. wide) and stipe silvery bluish white; whitish universal veil

Cortinarius alboviolaceus/ C. griseoviolaceus group

Cap ochraceous to grayish white in age. Stipe color same as cap. Universal veil leaves fibrillose remnants on the stipe that cover the pale violet surface. Flesh pallid to pale violet. In mixed forests but mainly associated with hardwoods, late summer to fall. Indistinct odor and taste. Harmless if eaten, but not recommended. Several ± similar species in NA, EU.

↓371b. Cap (< 4 in. wide) and stipe pale bluish lilac; strong unpleasant odor; sparse veil

Cortinarius camphoratus

Common. Cap ages yellow to yellow-brown from center. Partial veil leaves thin, fibrous ring zone. Flesh pale violet, dark ochraceous yellow at base. Odor of rotting meat, burnt flesh, goats. Taste unpleasant. Nontoxic.

371c. Cap (1½–5 in. wide), stipe, and veil pale lilac; strong fruity odor

Cortinarius traganus

Very common species best distinguished by flesh that is marbled rusty brown to yellow brown, more yellow at base of stipe. Odor of over-ripe pears, sickly sweet to some, taste sweetish to bitter. Reportedly difficult to digest. Not recommended.

372a. (361c) Colorless to grayish white universal veil *374*

372b. Bluish lilac or pale purple, glutinous universal veil *373a*

373a. (372b) Cap 1–4 in. wide, deep bluish lilac aging reddish brown

Cortinarius salor group

C. salor (EU, Asia) is similar to two yet unnamed Cascadia species. Young gills bluish lilac. Viscid stipe is whitish to pale lilac. Flesh-bluish at top of young stipes. Mixed woods, summer to fall. Indistinct odor, taste ± astringent. Edibility?

373b. Cap (1–3 in. wide); stipe with glutinous pale to dark purple veil

Cortinarius vanduzerensis   *Cortinarius seidliae*

The cap of *C. vanduzerensis* (Cascadia) is often corrugated at maturity, and the glutinous veil is dark purple. The cap of *C. seidliae* (Cascadia) is hygrophanous, margin initially pale and strongly radially wrinkled. The glutinous veil is faintly purple. *C. collinitus* (EU, NA) has violaceous gills when young. All are found in moderately wet, coniferous forests, late summer to fall. Odor ± sweet. Taste indistinct. Edibility? This group also has at least one unnamed species.

374. (372a) Cap (1–4 in. wide) glutinous; colorless veil

Cortinarius brunneoalbus   *Cortinarius mucosus*

C. brunneoalbus (Cascadia) cap hygrophanous, margin strongly radially wrinkled. Stipe white, viscid. Conifers fall. Odor faintly of honey. Taste? Edibility? *C. mucosus* (western NA, EU) slightly hygrophanous; young gills grayish white. Stipe viscid, white. Mixed woods, summer to fall. Odor ± fragrant, taste mild. Edibility?

375a. (361b) With purple-staining flesh *376a*

↓375b. Cap (1½–4 in. wide) yellow-brown to orange-brown; grayish white gills

Thaxterogaster multiformis
(= *Cortinarius multiformis*)

The dry white stipe ages brown, ±
marginate bulb. White, sparse veil. In
coniferous forests (often spruce), late
summer to fall in NA, EU. Odor faintly
sweet, taste sweet. Edibility?

↓375c. Cap (2–6 in. wide) yellow to red-brown; club-shaped stipe; ± green
corn–like odor

Phlegmacium superbum
(= *Cortinarius superbus*)

Cap and stipe undergo color change
from ± olive-yellow to red-brown.
Young gills straw-yellow. Stipe
often has viscid zone near base.
Under conifers in mountains, late
summer to fall in Cascadia. Taste not
distinctive. Edibility? *P. citrinifolium*
(= *C. citrinifolius*) is ochraceous, yellow
flesh and gills, fragrant odor. Edibility?

375d. Cap (1–4 in. wide) yellow-brown; white universal veil remnants on
cap and stipe

Phlegmacium variosimile group
(= *Cortinarius variosimilis* group)

Young gills may have a pale lilac tinge.
The white universal veil is strongly
developed, sheathing lower stipe;
cortina flocculose. Mixed coniferous
woods, fall in Cascadia. Two genetic
species. Indistinct odor and taste.
Edibility?

376a. (375a) Cap (1–4 in. wide) dark blue-violet, bruising purplish lilac

Thaxterogaster occidentalis
(= *Cortinarius occidentalis*)

Long known as *C. mutabilis.* Under firs, late summer to fall. Indistinct odor and taste. Edibility? *T. purpurascens* and *T. subpurpurascens* were also formerly in genus *Cortinarius.* These are known from Cascadia and EU. Edibility?

376b. Cap 1–3 in. wide, grayish aging ochraceous brown; bruises purple

Thaxterogaster porphyropus
(= *Cortinarius porphyropus*)

In *T. purpurascens* group. Young gills grayish blue. Veil sparse, violaceous tinged. Subalpine, mixed woods, fall in western NA, EU. Odor indistinct to like honey, taste mild. Not edible. Probably at least two similar species present in Cascadia.

377a. (361a) Young gills blood-red (page 288) *379a*

377b. Young gills rusty orange to bright orange *378a*

378a. (377b) Cap < 2 in. wide, yellowish olive aging reddish brown; stipe ± ochraceous

Cortinarius zakii

C. zakii (dull orange gills) and the similar *C. thiersii* (yellow gills) often pass for *C. croceus* (EU), a species that may not be present in Cascadia. Fruits under Douglas fir in fall. *C. thiersii* (CA, ID) fruits spring to summer. Both have a radish odor. Edibility? Known from Cascadia. Dye mushrooms.

378b. Cap (< 2 in. wide) red-brown, lighter margin; gills bright orange

Cortinarius subcroceofolius or *Cortinarius cinnamomeus*?

C. cinnamomeus is an eastern NA species and is probably the species in Cascadia. *Cortinarius subcroceofolius* was proposed as the species in Cascadia, but may prove to be a synonym of *C. cinnamomeus*. The cap can be hazel-brown to olive-brown. Stipe golden yellow to yellowish white. Cortina yellowish to red-brown. Under conifers, sometimes hardwoods, spring to fall. Odor weak, of radish. Edibility? A dye mushroom.

379a. (377a) Cap (< 2 in. wide), gills, flesh, and stipe blood-red

Cortinarius neosanguineus

Long called *C. sanguineus* (EU, eastern NA). Found in mixed woods and conifer stands, fall in Cascadia. Odor and taste mild to like radish. *C. birkebakii* (BC, WA, OR) is very similar, blackish red in the center and brownish red on cap margin with a buff-pink to red stipe. Edibility? Dye mushrooms.

379b. Cap (1–3½ in. wide) rich, dark red; gills deep red; stipe ± dull yellow

Cortinarius smithii

Formerly known as *C. phoeniceus* var. *occidentalis*. Cortina scanty, dull yellow to yellow-buff. Under conifers and in mixed woods, late summer to fall in western NA. Indistinct odor and taste. Edibility? The stipe base of red-capped varieties of *C. ominosus* (NA, EU) glows orange under UV. Dye mushrooms.

379c. Cap (< 2½ in. wide) yellow-brown to orange-brown; gills red; stipe yellow

Cortinarius ominosus

C. semisanguineus (EU) was a misapplied name. The lower stipe glows a rich orange-yellow under UV light. (*C. smithii* does not fluoresce.) Fleeting yellowish cortina. In conifer and mixed woods, summer to fall in NA, EU. Odor and taste ± of radish. Edibility? Dye mushrooms.

Note: *Cortinarius* species and species in the new segregate genera (see page 278) are far and away the most common mushrooms found under oak trees in Cascadia. The *Cortinarius* species associated with Oregon white oaks start fruiting in October after the initial fall rains and after a few light frosts. Hard freezing ends production. They are most abundant during November, and in a mild winter can persist past Christmas. They are often covered by oak leaves and can be hard to spot. Some oak groves rarely have any, while other oak groves produce many different species. The oak-associated species are in the genera *Cortinarius*, *Calonarius*, and *Phlegmacium*. The known oak-associated species in *Cortinarius* subgenus *Telamonia* have now been named. Most of the other oak-associated species remain unnamed. They will be added to MycoMatch as naming proceeds.

380a. (360a) Viscid caps; ± dry, often very thick stipes, with ± enlarged stipe base (genera *Phlegmacium* and *Calonarius*) (page 291) *384a*

380b. Smooth, dry caps, sometimes hygrophanous; ± white universal veil leaves bands on stipe; ± UV active (genus *Cortinarius*) *381a*

381a. Diameter of largest caps > 2 in.; stipe diameter ≥ pencil *383a*

381b. Diameter of largest caps < 1½ in.; stipe diameter < pencil *382a*

382a. (381b) Cap (< 1½ in. wide) reddish brown aging buff; pale brown flesh

Cortinarius albosericeus

Cap is silky dry, broadly umbonate, sometimes silvery gray-brown. Young gills pinkish cinnamon and have paler edges. The stipe is partially rooting. Under oaks with nearby ponderosa pine or Douglas fir, late fall in Cascadia. Odor and taste slightly fungoid. Edibility?

382b. Cap < 1¼ in. wide, brown aging ochraceous; vinaceous-purple gills

Cortinarius fragrantissimus

Hygrophanous cap, ± acute umbo. The stipe is hollow, pale pinkish buff with white mycelium. Under oaks, ± nearby Douglas fir or pine in Cascadia. Odor slightly sweet, taste mild. Edibility?

383a. (381a) Cap (< 2¼ in. wide) red-brown to cinnamon-colored; thin, white universal veil

Cortinarius vinaceogrisescens

Convex to flat, hygrophanous, silky cap, rarely any umbo. Stipe white, turning vinaceous-brown. In oak woods near Douglas fir, fall in Cascadia. Odor pleasant, slightly fragrant; taste mild. *C. roseobasilis* has a slight umbo. Edibility?

↓383b. Cap (< 2¼ in. wide) cinnamon-streaked with yellow-brown; white rhizomorphs

Cortinarius wahkiacus

Stipe < ½ in. wide and 3 in. long; universal veil leaves a thin whitish ring near stipe base. Under oaks near ponderosa pine, late fall in Cascadia. Odor musty to fishy, taste mild. Edibility?

↓383c. Cap < 2¼ in. wide, brown to red-brown aging pale yellowish white; universal veil almost sock-like

Cortinarius badioflavidus

Cap hygrophanous, broadly umbonate. Stipe < ½ in. wide and 3 in. long, tough, buff and yellow, watery red-brown in age, white universal veil. Under oaks and firs, fall to spring in Cascadia. Odor sharply fragrant to green corn–like, taste astringent. Edibility?

↓383d. Cap (2–6 in. wide) watery brown with whitish bloom; stipe clavate to bulbous

Cortinarius duboisensis

Large for the *Telamonia* subgenus. Cap hygrophanous, dark brown, fading to light buff, center lighter. Under oaks near ponderosa pine or grand fir, late fall in Cascadia. Odor mildly woodsy, taste fungoid. Edibility?

383e. Cap (< 5 in. wide) fibrillose, olive-green to reddish brown; bright yellow under UV

Cortinarius brunneofibrillosus   *Cortinarius cotoneus*

C. brunneofibrillosus (Cascadia) often has reddish brown discolorations. The odor is often of marjoram. Taste indistinct. Edibility? *C. cotoneus* (EU) is known from one collection in CA and one in Klickitat County, WA, but this could be a sister species.

384a. (380a) Cap lilac to lavender, quickly fading to white; lilac gills

Calonarius albidolilacinus
(= *Cortinarius albidolilacinus*)

Distinctive. Cap < 4 in. wide, ochraceous tones when young. Marginate stipe base. Under oaks in Cascadia. Flesh of stipe with lavender or grayish lilac tones, discolors yellowish. Odor and taste mild. Edibility?

↓384b. Cap ± orange-brown, turns wine-red with KOH; anise-like odor

Calonarius anetholens
(= *Cortinarius anetholens*)

Cap < 4 in. wide, multicolored yellow to orange-brown. Marginate stipe, pale greenish yellow with grayish green streaks. Flesh of stipe white, streaked grayish green. Taste mild. Edibility? In Cascadia.

↓384c. Fruitbodies pale with grayish, bluish, lilac, or yellowish tints

Phlegmacium aurescens
(= *Cortinarius aurescens*)

Cap (< 6 in. wide) turns distinctly yellow with KOH. Stipe white, slightly lavender in upper part, bulbous to marginate base. Gills ± pale lavender. Odor pleasant, taste fungoid. Edibility? In Cascadia.

↓384d. Cap (1½–4 in. wide) white with orange-buff tones; stains yellow

Phlegmacium argutum
(= *Cortinarius argutus*)

Cap dries readily, viscid nature seen from the debris "glued" on. Young gills, flesh, and stipe white. Under aspens and conifers as well as under oaks, late fall in EU, Cascadia. Faint odor. *P. albofragrans* (flowery to anise odor) is similar (see key lead 1a #2 [page 42]). *Calonarius xanthodryophilus* (leafy odor, page 321) has an abruptly bulbous stipe. Edibility?

↓384e. Cap (1–4 in. wide) viscid, dark bluish green; stipe grayish green

Phlegmacium glaucocephalus
(= *Cortinarius glaucocephalus*)

Stipe flesh marbled blue, stipe base a marginate bulb. Young gills dark blue. Under UV light cap dull red, stipe greenish yellow. Under conifers or under oaks, late fall in Cascadia. Odor musty, taste mild. Edibility?

↓384f. Cap (1–5 in. wide) viscid, bright yellow with orange-brown on disk

Calonarius amabilis
(= *Cortinarius amabilis*)

Stipe with a marginate bulb and yellow veil on rim of bulb and attached yellow basal mycelium. Young gills lilac-beige. Associated with Oregon white oak, late fall in Cascadia. Indistinct odor and taste. Edibility?

384g. Cap (2–6 in. wide) yellowish with ochre-brown; young gills yellowish

Calonarius vellingae
(= *Cortinarius vellingae*)

Very large. Stipe marginate, white; with white, yellow, pinkish, to reddish basal mycelium. Veil thin, yellow. Under live oak, tanoak, and Oregon white oak in Cascadia. Odor and taste mild. Edibility?

385a. (175a) Gills neither decurrent nor free (page 299) *396a*

↓385b. Gills decurrent (genera *Gomphidius* and *Chroogomphus*) (page 297) *392a*

385c. Gills free; cap may break from stipe in ball-and-socket fashion *386a*

386a. (385c) Spore print chocolate-brown; always with a partial veil (*Agaricus* spp.) *387a*

↓386b. Spore print black; scaly, bullet-shaped cap dissolves into black ink

Coprinus comatus

A very distinctive, choice edible mushroom, 3–8 in.tall, cap always much longer than wide. The cap dissolves from the bottom up, turning into black ink. The cap pops from the stipe and only the cap is eaten. Needs to be eaten the day of harvest, but can be kept longer if submerged in water. Microwaving for ± 30 seconds also supposedly slows decomposition. Can grow in large clusters on lawns, roadsides, trails, hard-packed ground, spring to fall in NA, EU, Asia, North Africa, Australia.

386c. Not as above: gills free, spore print not any light color, and not with the features in 386a or 386b (page 311) *413a*

387a. (386a) Caps < 2 in. wide; sweet almond scent; KOH yellow (page 296) *391a*

↓387b. Yellow staining, ± bright; ± creosote odor; KOH yellow (page 295) *390a*

↓387c. Weak yellow staining; sweet almond odor; KOH yellow (page 294) *389a*

387d. No staining; odor pleasant, not of sweet almond; KOH negative or green *388a*

388a. (387d) No staining; mild odor; KOH negative

Agaricus porphyrocephalus

Cap (2–5 in. wide) white or appressed brown-scaly; young gills pink; ring scanty, with veil remnants mainly on cap margin. In grass, pastures, barnyards, spring to winter in cool, moist weather. Odor and taste pleasant, hint of almonds. Choice edible. *A. campestris* (NA, EU, North Africa) is a look-alike.

388b. No staining; mild odor; KOH green; cap 2–8 in. wide

Agaricus subrutilescens

Slow green-staining reaction. The wine-red cap and woolly stipe are distinctive. Under hardwoods and conifers, fall in Cascadia. Odor mild to slightly fruity, taste pleasant. Edible for most people. Avoid the similar *A. hondensis* with a smooth stipe, creosote-like odor, yellow staining.

389a. (387c) Cap (< 16 in. wide) with tawny fibrils; young gills pallid

Agaricus augustus (atypically huge specimen at left)

Distinguished by large size, yellow-staining cap, and shaggy lower stipe. Usually along roads, paths, near compost heaps, or disturbed ground, spring to fall, often in the hot-dry heat of August in NA, EU, Asia, North Africa. Odor and taste pleasant, of anise or almonds. Highly prized, distinctive edible. Any pleasant, sweet-smelling *Agaricus* species is edible (for most people). Any *Agaricus* species with a phenolic, creosote-like, or unpleasant odor fresh or when cooking is likely to make you sick if you eat it.

389b. Cap (2–5 in. wide) white, bruising ± yellow; veil thin, white ring; stipe ± bulbous

Agaricus sylvicola group

A. abruptibulbus is shown in photo. *A. moronii* is similar but larger, bruises more strongly yellow than orangish, and the stipe is thicker, not bulbous. Both grow with conifers, spring to fall in NA. Both have a pleasant odor and taste like almond or anise. Edible for most people.

390a. (387b) Cap (2–7 in. wide) with dark grayish brown to blackish fibrils, scales

Agaricus deardorffensis/
A. buckmacadooi

Both long known as *A. moelleri* (EU) and *A. praeclaresquamosus* (EU), these hard-to-distinguish species are near conifers, summer to fall in western NA. The odor is phenolic or of wet asphalt. Causes significant GI distress if eaten.

↓390b. Cap (2–8 in. wide) pink-tinged, fibrillose, darkening in age

Agaricus hondensis   *Agaricus subrufescentoides*

A. hondensis has a sheathing ring. The interior flesh at stipe base is yellowish, often slowly discoloring pinkish. The stipe is smooth, brownish in age or after handling, typically bulbous. *A. subrufescentoides* is a darker, often slenderer, non-staining look-alike. In low-elevation coniferous forests, sometimes under redwoods, fall in Cascadia. Odor indistinct to phenolic. Causes GI distress if eaten.

390c. Cap (2–8 in. wide) white to gray, quickly staining bright yellow

Agaricus xanthodermus

The rapid yellow staining when bruised is distinctive. Membranous veil with felt-like ring. Under trees in lawns, woods, and pastures, often with ponderosa pine, summer to fall in NA, EU, North Africa. Odor inky to phenolic. Causes very severe GI distress if eaten.

391a. (387a) Cap (1–3 in. wide) pinkish, aging brown, no umbo; veil persistent, thin

Agaricus diminutivus

Little *Agaricus* species (subgenus *Minores*) are difficult to distinguish. Some authors report purplish fibrils in the cap of *A. diminutivus* (= *A. purpurellus*?). Under conifers, summer to fall in Cascadia. Odor of almonds, anise. Edible but insubstantial.

↓391b. Cap (1–3 in. wide) wine-red in center, light gray margin, upturned ring

Agaricus 'purpurellus'

Believed to be a color variant of *A. diminutivus* but likely to be one or more distinct members of subgenus *Minores*. Gills initially pale gray-vinaceous. Stipe club-shaped to bulbous, slightly flocculose base. Odor and taste sweet, almond-like. Edible, but insubstantial. In Cascadia.

↓391c. Cap < 2 in. wide, appressed pinkish red fibrils aging gray-brown; bruises yellow

Agaricus 'micromegethus'

A. 'diminutivus' and *A. 'purpurellus'* have a cottony lower stipe, while *A. 'micromegethus'* and *A. 'comptulus'* are ± smooth. Found in lawns, pastures, and grassy areas, fall in NA, EU. Odor and taste almond-like. Edible but rather small. Several different members of subgenus *Minores* will key out here; DNA will be needed to distinguish and eventually name them.

391d. Cap 1–3 in. wide, appressed fibrils pinkish aging brownish

Agaricus kerriganii

Conifer associate, similar to *A. diminutivus*. Confirmed by DNA to be *A. kerriganii*. Sparse, cottony scales on lower stipe. Stipe and cap bruise initially yellow turning orangish. Ring soon appressed to stipe and ± disappearing. Odor and taste of anise, almonds. Edible. Range NA, EU, North Africa.

392a. (385b) Flesh white; typically yellow at stipe base (page 298) *394a*

392b. Flesh yellow-orange; ± yellow at stipe base *393a*

393a. (392b) Cap (2–5 in. wide) viscid, ± pointed umbo, ochraceous orange to ± red

Chroogomphus ochraceus group

C. rutilus was a misapplied EU name. The stipe is dry to moist, orange-buff with a reddish flush and fine fibrils below faint ring zone from fibrillose veil. Flesh ± light salmon, yellowish at stipe base. Appears parasitic on *Suillus fuscotomentosus* mycelium under ponderosa pines, late summer to fall in NA. Indistinct odor, taste indistinct but pleasant. Edible. An unnamed *C. vinicolor* look-alike is very similar, distinguished by a convex cap and pointed umbo.

↓393b. Cap (2–6 in. wide) dry to ± viscid; stocky stipe ≤ 2 in. wide

Chroogomphus pseudovinicolor

Cap orange-buff, mottled with reddish, felty material. Firm stipe, zones of reddish, felty material below scanty ring zone. Fibrillose veil. Flesh dull orange. Under Douglas fir and ponderosa pine above 2,500 ft., summer to fall in Cascadia. Indistinct odor, pleasant taste. Edible.

393c. Cap (1–4 in. wide) dry, woolly, dull orange; stipe ± equal to tapered

Chroogomphus tomentosus

Distinctive. Common at all elevations where hemlocks are present (believed parasitic on *'Aureoboletus' mirabilis* mycelium), summer to winter in Cascadia and Japan. Indistinct odor and taste. Turns wine-red when cooked. Mediocre edible.

394a. (392a) Flesh at base of stipe yellow; glutinous veil present *395a*

↓394b. Flesh at base of stipe usually not yellow; no veil; under larch

Gomphidius maculatus

Cap (< 4½ in. wide) light cinnamon-colored to reddish brown, bruises black. Stipe ± dry, lower two-thirds with dark fibrils. Flesh bruises red, then black, or black directly. Mild odor and taste. Edible, but peel cap first. Range NA, EU.

394c. Cap (< 2½ in. wide) grayish to vinaceous-buff

Gomphidius smithii

Very glutinous cap. Stipe widest in middle, may be yellow at extreme tip. Veil colorless gluten over white fibrillose layer. Under lodgepole pine and Douglas fir, summer to fall in Cascadia. Parasitic on *Suillus lakei* mycelium. Odor and taste mild. Edible, little flavor.

395a. (394a) Cap < 7 in. wide, salmon-buff aging vinaceous gray; usually cespitose

Gomphidius oregonensis

Cap stained smoky black in age. Veil colorless gluten over white fibrillose layer. Widespread under Douglas fir and other conifers, summer to fall in Cascadia. Parasitic on *Suillus caerulescens* mycelium. Indistinct odor and taste. Edible, peel off slime first.

395b. Cap (< 3 in. wide) dull, bright pink, or rosy red

Gomphidius subroseus

Like all *Gomphidius* species, the cap skin is a thick, peelable, glutinous pellicle. The flesh is white, yellow in lower quarter of stipe, sometimes pink at extreme base. Common under conifers, summer to winter (parasitic on *Suillus lakei*?) in NA. Mild odor and taste. Bland edible.

396a. (385a) Cap 2–15 in. wide, wine-red aging tan; cultivated areas

Stropharia rugosoannulata

Cap is slightly viscid to dry. White stipe discolors brownish, club-shaped base often with copious white mycelial threads. Gills initially whitish. Ring grooved or segmented. Often in rich ground with wood chips, never in the woods, spring to fall in NA, EU, Asia. Pleasant odor and taste. Edible and good especially when young. Very distinctive.

↓396b. Cap (2–6 in. wide) viscid, ± red-brown; flocculose lower stipe

Stropharia hornemannii

Cap fades to yellow-brown when old. Veil leaves hanging particles on cap margin and lower stipe and leaves a ± persistent ring that is thin and striate on the top side. Under conifers or on rotting wood, summer to fall in NA, EU. Odor none to unpleasant, taste disagreeable. Not a tasty edible.

↓396c. Cap (2–6 in. wide) viscid, yellow; flocculose veil remnants on margin

Stropharia ambigua

Very elegant looking and very common in all types of woods at all elevations, sometimes spring, usually fall, in Cascadia. Somewhat fragile. Gills initially pale gray. Odor and taste faint to of old leaves. Not a tasty edible.

396d. Not as above: spores chocolate brown to black and not covered in 396a–c (page 300) *397a*

397a. (396d) Cap ± fragile, ± fringe on margin; stipe ± fragile (*Psathyrella* spp.) (page 307) *408a*

↓397b. Cap bald, ± fringe on cap margin; gills mottled (*Panaeolus* spp.) (page 306) *405a*

↓397c. Cap not viscid, ± fringed margin; gills not mottled (*Hypholoma* spp.) (page 305) *404a*

397d. Cap viscid, ± fringed; stipe ± flocculose (*Stropharia* spp. and allies) *398a*

398a. (397d) Cap hygrophanous, peelable; stipe bald (*Psilocybe* spp. and allies) *400a*

↓398b. Cap bald, margin ± fringed, ± yellow to ± brown *399a*

↓398c. Cap (1–3 in. wide) bald, margin ± fringed, greenish blue; stipe white

Stropharia aff. *aeruginosa*

Cap blue, aging greenish yellow, margin with veil remnants. Stipe with cottony scales below ring. In rich soil, woody debris, grass, woods, in low elevations, spring to fall. A look-alike to S. *aeruginosa* (EU). Indistinct odor and taste. Possible GI distress if consumed. Not hallucinogenic.

↓398d. Cap < 2 in. wide, removable fibrils, orangish; lower stipe cottony

Leratiomyces 'squamosus' var. *squamosus*

Formerly known as *Psilocybe squamosus*. Cap dull yellow–ochre to tawny, ± hygrophanous. Gills initially pallid. White membranous veil. Ring with striations on top. Stipe brownish, covered with fibrillose white scales. In mixed conifer and alder woods, on woodchips, late summer to fall in NA, EU, North Africa. Indistinct odor and taste. No psilocybin. Edibility? Possibly actually an unnamed look-alike.

398e. Cap < 2½ in. wide, removable fibrils, orangish red to brick-red

Leratiomyces ceres

Stipe with small brown scales when young, aging reddish brown. Global distribution. On deciduous twigs, branches, and wood chips, late summer to fall. Indistinct odor and taste. No psilocybin. Edibility?

399a. (398b) Cap diameter typically 2–3½ in.; near alders or cottonwoods

Leratiomyces riparius/L. percevalii

Distinguished from *S. ambigua* by a veil that leaves remnants on the cap margin, rather than hanging from the margin, and a less flocculose lower stipe. In seepage areas and along streams in western NA (*L. riparius*). Odor and taste mild. Edibility? May be *L. percevalii* (EU).

↓399b. Cap (≤ 2 in. wide) convex, ± straw-colored; stipe viscid below delicate ring

Protostropharia semiglobata group

Formerly known as *Psilocybe semiglobata*. Cap not hygrophanous, separable pellicle. Gills initially grayish. On manure, rich soil, manured grass, spring to fall in NA, EU, Asia, North Africa, Australia. Odor weakly farinaceous or spicy; taste sweet to insipid. Mediocre edible.

399c. Cap ≤ 2 in. wide, convex, flat in age, ± viscid, creamy to gold-brown; persistent ring

Stropharia coronilla

Shorter than *P. semiglobata* with a more prominent ring. Young gills pallid, gill margins remain whitish. In grassy areas, late summer to winter in NA, EU, Asia, North Africa, Australia. Odor faintly pungent, taste not distinctive. Possibly somewhat toxic. No psilocybin.

400a. (398a) Bruises blue when bruised; cap bell-shaped ± with a nipple; on the ground in boggy pastures (page 303) *402a*

400b. Not bruising blue, cap convex to broadly umbonate (page 302) *401a*

401a. (400b) Cap (< 1 in. wide) yellowish to red-brown, slight umbo; young gills whitish

Deconica subviscida

Formerly known as *Psilocybe subviscida* and *Deconica (Psilocybe) graminicola*. Cap hygrophanous, ± peelable. Veil thin, ring ± absent. Stipe lightly fibrillose below. On well-manured ground or conifer debris, spring to fall in NA, EU, Australia. Odor and taste slightly fungoid. No psilocybin. Edibility?

↓401b. Cap (< 1 in. wide) shades of brown, not umbonate; young gills gray-brown

Deconica coprophila

Formerly known as *Psilocybe coprophila*. Almost always on cow or horse manure, year-round. Global distribution. Odor and taste slightly fungoid. Edibility? No psilocybin or psilocin.

↓401c. Cap < 1 in. wide, red-brown aging straw-colored; young gills pallid; pale to reddish stipe

Deconica aff. inquilina

Cap margin translucent-striate. The flesh is thin and pliant. Grows directly on rotting grass and twigs, attached with white mycelial felt at stipe base. Several species associated with moss have also been known as *D. inquilina*. Global? Indistinct odor and taste. Edible?

401d. Cap < 1 in. wide, reddish brown aging straw-colored; pallid young gills with paler edges

Deconica montana

Formerly known as *Psilocybe montana*. Cap viscid from a peelable pellicle, translucent-striate. Stipe ± cap color. Presumably on dead root masses of mosses, spring to fall, circumglobally. Odor and taste slightly fungal. Edibility? No psilocybin or psilocin.

Note: All *Psilocybe* species are at low elevations west of the Cascades and Sierras. *Psilocybe* species can be confused with deadly *Galerina* and *Pholiotina* species.

402a. (400a) Typically found growing in grass on school grounds, sports fields, prison exercise yards in wood chips, bark mulch

Psilocybe stuntzii

Cap (< 2 in. wide) striate, peelable skin, deep chestnut-brown when moist, yellowish buff on moisture loss. Thin, membranous bluing ring on yellowish to brown stipe. Fall to early winter, rarely spring, in Cascadia. Odor and taste farinaceous when young. Total psilocybin and psilocin variable, 0–0.5 percent. Mildly hallucinogenic.

↓402b. In wet pastures with rushes and sedges

Psilocybe semilanceata

The hygrophanous cap (< 1 in. wide) is sharply conic with a small nipple, changing from chestnut-brown to yellowish on drying. The stipe is slender. Veil rarely seen, no ring. Spring to winter, mainly fall, in many parts of the world. Contains psilocybin (± 1 percent) but no psilocin, thus rarely bruises blue. Odor and taste mildly fungoid.

↓402c. In woods on conifer debris, wood chips; rarely bruises blue

Psilocybe pelliculosa

The hygrophanous cap is conic to bell-shaped without a nipple. Chestnut-brown when moist, tan when dry. Cobwebby white veil when young, ± obscure ring, leaves white fibrils on lower stipe. Odor and taste ± farinaceous. Psilocybin 0–0.4 percent. Mildly hallucinogenic. Distribution NA, EU.

402d. Associated with woody debris, always strongly bluing (page 304) *403a*

403a. (402d) Cap (1–4 in. wide), chestnut-brown, hygrophanous, peelable, not wavy; strong blue bruising

Psilocybe azurescens

In beach grass–covered sand dunes September to January, and on beds of wood chips. Indistinct odor, very bitter taste. Psilocybin + psilocin exceeds 2 percent by dry weight with 0.3–0.4 percent baeocystin. Can cause semi-paralysis lasting about an hour.

↓403b. Cap with a ± peelable pellicle < 2 in. wide, ± chestnut aging tan, margin ± wavy; strong blue bruising

Psilocybe cyanescens

Fibrillose veil may leave faint ring. On wood chips, sawdust, bark mulched beds in Cascadia, EU. Often densely gregarious. Odor and taste ± farinaceous. Potent! Psilocybin + psilocin can exceed 2 percent.

403c. Cap < 1½ in. wide; strong blue bruising; farinaceous odor and taste

Psilocybe cyanofibrillosa *Psilocybe allenii* *Psilocybe baeocystis*

These three species may have originated in coastal river estuaries. Found now on wood mulch and easily spread to new areas. *P. cyanofibrillosa* about 0.3 percent psilocybin + psilocin. *P. allenii* can exceed 2 percent. *P. baeocystis* is about 1 percent total.

404a. (397c) Cap (1–3 in. wide) tawny to ± orangish; young gills grayish

Hypholoma capnoides

Formerly known as *Naematoloma capnoides*. Cap margin ± with veil remnants. Stipe with ± obscure fibrillose zone. Clustered on rotting conifers, late summer to winter (gray-gilled wood lover) NA, EU, Asia. Odor and taste fungoid. Edible, but mediocre taste. Has a ± dangerous look-alike, *H. fasciculare*.

↓404b. Cap (1–4 in. wide) greenish to orange-yellow; young gills ± greenish

Hypholoma fasciculare (both images)

Formerly *Naematoloma fasciculare* (green-gilled wood lover). Cap ± with veil remnants. Stipe ± obscure fibrillose zone. Clustered on rotting conifers or hardwoods, year-round, common in NA, EU, Asia, North Africa. Odor mild, taste slowly bitter. Poisonous (deadly in Asia). At maturity, *H. capnoides* and *H. fasciculare* are hard to distinguish.

↓404c. Cap (< 1½ in. wide) tawny to yellowish; slim stipe ± fibrillose patches

Hypholoma dispersum

Cap margin often with veil remnants from fleeting cobwebby veil. Only slightly hygrophanous. Under conifers, summer to fall, often abundant in NA, EU. *Bogbodia uda* (*H. udum*, page 321) is in bogs. Odor (both) indistinct, taste ± unpleasant. Edibility?

404d. Cap < 1¼ in. wide, orange-brown aging tawny; shiny cartilaginous stipe

Mythicomyces corneipes

Combines features found in many genera. The slim, somewhat horny, round stipe and young whitish gills are distinctive. In mosses, around bogs near alders and conifers in NA, EU. Indistinct odor and taste. Edibility?

405a. (397b) In grass, on manure or rich ground; with partial veil *407a*

405b. In grass, on manure or rich ground; partial veil absent *406a*

406a. (405b) Cap < 1½ in. wide, chestnut-brown aging tan; margin fades last; in grass

Panaeolus foenisecii

Cap hygrophanous with margin fading last. The dark purple-brown spore print (versus the normal black for *Panaeolus* spp.) caused former placement of this species in genus *Panaeolina*. In grass, spring to fall in NA, EU. Odor fungoid, taste mild. Harmless in western NA. Elsewhere may contain a small amount of psilocybin.

↓406b. Cap < 1½ in. wide, dark cinnamon aging tan; margin fades first; in grass

Panaeolus castaneifolius
(= Panaeolina castaneifolia)

Hygrophanous cap remains darker in center than on margin as it dries. Cap margin often wrinkled. Stipe brittle to cartilaginous. Black spore print. Uncommon. In grassy fields, summer to fall in NA. Odor and taste mild to unpleasant. Some collections reportedly contain small amounts of psilocybin but no psilocin, not bluing. Possibly mildly hallucinogenic. The one available DNA sequence turned out to be same as for *P. foenisecii.*

406c. Cap < 2 in. wide, cinnamon aging tan; margin fades last; on grass or dung

Panaeolus subbalteatus group

A look-alike to *P. foenisecii,* but distinguished by black spore print. On dung (especially horse) and in fertile grass, spring to fall, widespread globally. Indistinct odor and taste. Contains some psilocybin, a trace of baeocystin, and no psilocin (thus not bluing). Along with *P. castaneifolius,* the only *Panaeolus* species in Cascadia with psilocybin. Mildly hallucinogenic. Several look-alikes may be in Cascadia.

407a. (405a) Partial veil inconspicuous to absent; on dung or rich grass

Panaeolus acuminatus

Cap (< 1¼ in. wide) bell-shaped, expanding to umbonate, hygrophanous (chestnut aging tan, margin usually fading last), viscid, striate when moist. Stipe ± cap color or red-brown. Found spring to early winter in Cascadia, EU. Indistinct odor and taste. Edibility?

↓407b. Partial veil leaving hanging remnants on cap margin; in grass or dung

Panaeolus papilionaceus group

Current name for *P. campanulatus* and *P. sphinctrinus*. *P. retirugis* (very wrinkled cap) is genetically distinct. Cap (< 1¼ in. wide) never expanding to plane, hygrophanous dark grayish to whitish. No ring on stipe. Usually on dung, spring to fall circumglobally. Odor and taste weakly fungoid. Edibility? Does not contain psilocybin.

407c. Veil may leave fragile membranous ring; cap (1–3 in. wide) white to buff; on dung

Panaeolus semiovatus

The cap and is viscid, often with hanging veil remnants on margin. On dung, especially horse droppings, summer to fall in NA, EU, Asia, North Africa, Australia. Indistinct odor and taste. Not found to contain psilocybin. Not poisonous.

408a. (397a) Cap not scaly, not parasitic on *Coprinus comatus* (page 309) *410a*

↓408b. Cap and stipe scaly (page 308) *409a*

408c. Cap (1–3 in. wide) white, silky-fibrillose on old *Coprinus comatus*

Psathyrella epimyces

Distinctive. White veil leaves a
volva-like ring at base of white stipe.
Indistinct odor and taste. Edibility?
Range NA.

409a. (408b) Cap (1–4 in. wide) whitish to ochraceous, with grayish
brown scales

Lacrymaria 'lacrymabunda'

Formerly known as *Psathyrella velutina*. The
cap will expand to ± flat in age and the scales
easily rub off. A fibrillose veil leaves a faint ring
zone. Stipe sometimes stains yellowish. In
cespitose clusters under hardwoods in the fall
with two species in Cascadia, but so far not
L. lacrymabunda (EU). Odor and taste fungoid.
Edibility not reported but probably nontoxic.
Best distinguished from *L. rigidipes* (NA) by
spore size and shape.

409b. Cap (1–5 in. wide) finely scaly, tawny brown; young gills brownish red

Lacrymaria 'rigidipes'/L. 'velutina'

Cap is dry. White fibrillose veil does
not leave a ring on stipe. Stipe slightly
paler than cap, fibrillose scaly. Seen in
damp, grassy areas and on roadsides.
Fruits gregariously late summer to fall
in NA. Cascadia species names and
total number of species unclear. Odor
not noticeable. Taste mild. Edibility?

410a. (408a) Veil leaves whitish hairs on cap margin or hanging (page 310) *412a*

410b. Veil leaves distinct hairs on cap and/or a ring zone *411a*

411a. (410b) Small cap (< 1 in. wide) with white hairs; on dung or rich ground

Psathyrella hirta

Cap dark rusty brown, fading to dingy buff on moisture loss. Veil leaves fibrillose fibers on lower two-thirds of white stipe, soon disappearing. Young gills pallid, white margin in age. Summer to fall in NA, EU. Odor and taste mild. Edibility?

↓411b. Cap (< 2 in. wide) with white cottony-fibrillose hairs; on decaying wood

Psathyrella piluliformis group

Formerly known as *P. hydrophila*. Hygrophanous cap reddish brown when moist to yellow-brown when dry. Veil somewhat membranous, leaving cap margin remnants and obscure hairs on white stipe. On decaying hardwoods, spring to fall, two look-alike clades in NA, EU, Asia, North Africa. Indistinct odor and taste. Not recommended.

411c. Cap (1–4 in. wide) fragile but persistent ring; under conifers, alder

Psathyrella longistriata group

Appears to be more than six *Psathyrella* species in Cascadia with a persistent (though fragile) ring on stipe. The hygrophanous cap is ± reddish brown when moist and ± pallid when dry. A slightly wrinkled cap margin is common. Young gills pale buff. Fall, sometimes spring, common in NA. Indistinct odor and taste. Edibility?

412a. (410a) Cap (1–4 in. wide) brownish to yellowish when moist, whitish when dry

Candolleomyces candolleanus (= Psathyrella candolleana)

A highly variably colored, common species. Veil leaves some hanging remnants on cap margin, sometimes a brief fibrillose ring on stipe. Young gills pallid. Often with cottonwoods, spring to fall in NA, EU. Mild odor and taste. Edible.

↓412b. Cap (< 1½ in. wide) greenish cinnamon when moist, reddish brown when dry

Psathyrella abieticola

Hint of a veil leaves a thin, whitish margin on the cap. Stipe white, lacking many fibrils. Gregarious under true firs and spruce, at least in fall in Cascadia. Indistinct odor and taste. Edibility? Many ± similar fragile Psathyrella species.

↓412c. Cap (1–3 in. wide) rusty brown when moist, wood-brown when dry

Psathyrella spadiceogrisea

Cap smooth to finely wrinkled. Young gills pallid brownish. Veil may leave a few fibrillose remnants on cap and stipe, soon disappearing. In muck on edge of boggy ground, spring in NA, EU, Asia, North Africa. The odor is fungoid, taste mild. Edibility?

412d. Cap (1–4 in. wide) gray; young gills pallid; stipe ± ½ in. wide, whitish

Coprinopsis uliginicola

This oddball mushroom was once in genus *Psathyrella*, but it looks more like a *Tricholoma* species than anything else, except for the dark vinaceous-brown spore print. Found in a wet oak woodland, fall in western NA. Odor, and taste indistinct. Edible?

413a. (386c) Cap lead-gray to brownish, ≤ 2 in. wide; stipe ± ⅓ in. wide; cap turns to ink at maturity

Coprinopsis atramentaria group

Coprinopsis striata

Coprinopsis cf. *romagnesiana*

Coprinopsis atramentaria

C. striata (Cascadia) and *C. acuminata* (widespread) have narrow umbonate caps with brown striations, while *C. atramentaria* (widespread) has a broader, non-umbonate cap. *C. romagnesiana* (widespread, uncommon) is scaly below a faint ring zone. All contain coprine, a carcinogen that causes Antabuse-like symptoms (nausea, vomiting, dizziness, headache) if alcohol is consumed after eating. Not recommended.

↓413b. Small, delicate; crowded, near wood; cap not turning to ink

Coprinellus disseminatus

Cap < ½ in. wide, oval expanding to bell-shaped, pallid to buff with honey-brown center. Very thin and fragile. No veil. In woods or grassy areas near old stumps, spring to fall in NA, EU. No odor, taste mild. Edible?

↓413c. Cap (< 2 in. wide) tan to greenish yellow covered with glistening particles

Coprinellus micaceus

The glistening, ± round particles are the remains of the universal veil and readily wash off. The gills ± turn to ink. Grows in sometimes massive clumps on wood or near woody debris, spring to fall in NA, EU, Asia, North Africa, Australia. Odor and taste mild. Edible but bland.

↓413d. Cap < 2 in. wide, honey-brown blackish in age; whitish bran-like particles

Coprinellus flocculosus?

Several fragile species have caps covered with bran-like to felty particles (remains of a universal veil) that readily wash off. These include *C. flocculosus* and the *C. domesticus* group. The whitish stipe can look finely sanded or smooth. Tends to grow singly or in small clusters on wood chips in cool, wet weather in NA, EU. Indistinct odor and taste. Harmless but of no interest as an edible.

↓413e. Cap (< 2 in. wide) grayish brown, densely covered with grayish hairs

Coprinopsis lagopus group

Gills turn black and inky or wither in dry weather. Single or in troops on woody debris, dung, compost piles in NA, EU, Asia, North Africa. Indistinct odor and taste. Half a dozen flavorless species.

413f. Cap (< 1½ in. wide) snow-white from a mealy, powdery veil; on dung

Coprinopsis nivea

White gills turn black but retain a white edge. Stipe white, brownish toward base but with white veil flecks. On dung of horses and cows in NA, EU, Asia, Australia. Odor and taste indistinct. Edibility?

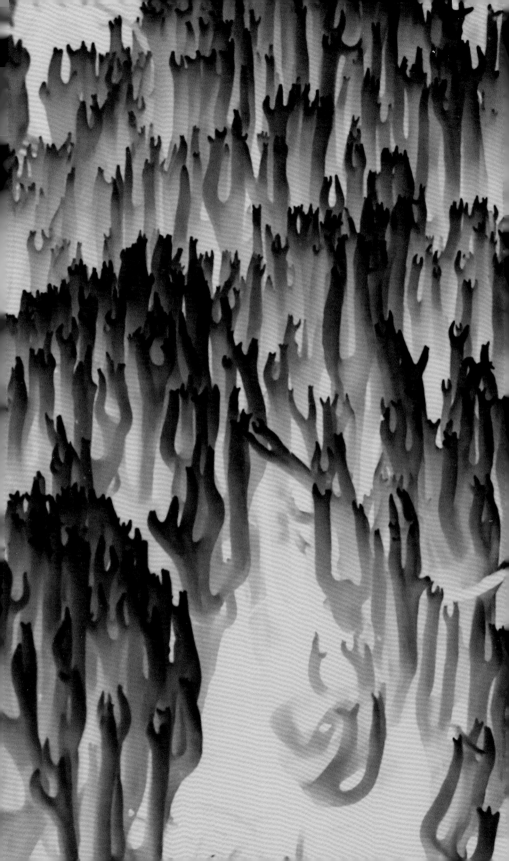

Addendum

Since the publication of this book's first edition, I have found and photographed more than forty species that were only mentioned in the comments. The entries before the species names (for example, 46b #1, 46b #2, 46b #3, 50a) on the following pages indicate the key lead where the species appears in comments 46b and 50a. The number is the order in which the species are mentioned. Full descriptions of the named species can also be found in MycoMatch.

46b.

#1 *Gyromitra mcknightii* (in *G. ancilis/perlata* group)

#2 *Gyromitra leucoxantha* PNW01 = *Gyromitra persicula* (in press)

#3 *Disciotis* PNW01 (in *D. venosa* group)

#4 50a *Daleomyces bicolor*

51d.

Plectania milleri PNW01

52c.

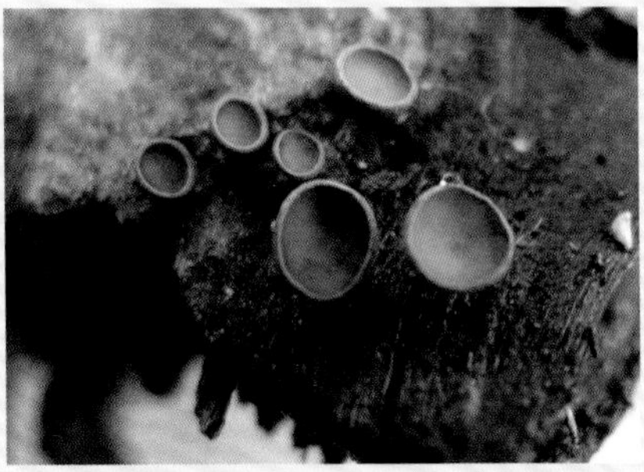

Chlorociboria aeruginosa

55b.

#1 *Cheilymenia theleboloides*

#2 *Coprotus* cf. *ochraceus*

57c.

Pithya cupressina

76b.

Stereum ochraceoflavum

90e.

Porodaedalea chrysoloma
(= *Phellinus chrysoloma*)

91d.

Funalia 'gallica' IN01
(= *Coriolopsis 'gallica'* IN01)

95b.

Peniophora aurantiaca

96a.

Rhizina undulata

102c.

Tuber luomae

105.

Agaricus inaperatus

111a.

Lycoperdon 'vernimontanum'

124c.

Ramaria subviolacea

127a.

Ramaria cystidiophora var.
maculans

128b.

Ramaria magnipes PNW01

136a.

Ramaria flavigelatinosa var.
megalospora

136b.

Ramaria aurantiisiccescens

138a.

Ramaria cyaneigranosa var. *elongata*

161c.

Tremella mesenterella

194c.

Leucoagaricus fuliginescens PNW05

206c.

Collybia diatreta

222b.

Tricholoma intermedium

224a

Gymnopus spongiosus group

232a.

Collybia cirrhata

236a.

Strobilurus albipilatus

252.

Gymnopus androsaceus

264d.

#1 *Hygrocybe cantharellus*

264d.

#2 *Hygrocybe cantharellus*

264d.

#3 *Hygrocybe parvula* PNW01

301c.

Lactarius 'alpinus'

302d.

Lactarius cordovaensis
(= *cascadensis*)

303c.

Lactarius atrobadius

311b.

Sarcomyxa PNW01

314b.

#1 *Hohenbuehelia 'angustata'*

#2 *Pleurotus ostreatus*

320

329c.

#1 *Paxillus cuprinus*

#2 *Paxillus involutus* group

362a.

#1 *Russula chlorineolens*

#2 *Russula suboculata*

384d.

404c.

Calonarius xanthodryophilus
(= *Cortinarius xanthodryophilus*)

Bogbodia uda (= *Hypholoma udum*)

Glossary

aborted: partly formed due to arrested development

adnate: attached broadly

aff.: genetically close to

agarics: gilled mushrooms

agglutinated: appears glued together, usually of fibrils

alkaline odor: smells like ammonia, bleach, or chlorine

alveoli: pit-like depressions (as, for example, on the head of a morel)

amyloid: staining blue, blue-gray, gray, or nearly black when exposed to iodine solutions (Melzer's Lugol's reagent)

anamorph: imperfect/asexual state of a fungus

annual: fruits for one year and dies

annulus: a ring on the stipe from remains of partial veil

annulate: when a partial veil leaves a ring on the stipe

apothecium, apothecia: an open cup-shaped ascomycete fruitbody

appressed: flattened onto a surface

ascocarp: ascomycete fruitbody (= ascoma)

ascomycete: a member of the Ascomycota

Ascomycota: a phylum of fungi in which the production of sexual spores occurs within an ascus

ascus, asci: a sac-like cell where sexual Ascomycota spores form

avellaneous: a color that ranges from drab to pale grayish brown to hazel

basal, basally: nearest to the point of attachment

basidiolichen: a lichen associated with a gilled fungus

concolorous: having the same color

confluent: merging

conic, conical: cone-shaped

conifer: a generally evergreen, cone-bearing tree

contorted: twisted or bent out of its normal shape

convex: bulging upward or outward in a rounded manner without any depressions

convoluted: intricately folded or twisted

cortex: rind, outer layer

cortinate: with a cobweb-type partial veil from the edge of the cap to the stipe

crenate, crenulate: scalloped, finely scalloped

cupulate: cup-shaped

decurrent: gills attach partway down the stipe

dextrinoid: turns reddish brown on application of Melzer's or Lugol's reagent

discoid: shaped like a shallow disc

distal: located away from the point of origin or attachment

duff: plant litter making up the surface layer of forest soil

eccentric: away from the center, off-center

echinate: covered with pointed spines

echinulate: covered with small, pointed spines

effuse, effused: spread out, expanded

elata clade: a genetic subgroup of black morels evolved from a common ancestor

ellipsoid, ellipsoidal, elliptic, elliptical: elongated with rounded ends and curved sides

ephemeral: lasting for a very short time

epigeous: fruitbodies that grow on or above ground; compare hypogeous

epiphyte: growing on the outside of a plant, but not as a parasite

epithet: the second half of a species' scientific name

equal: of the same diameter throughout

erumpent: breaking through or bursting out

esculenta clade: a genetic subgroup of yellow morels evolved from a common ancestor

evanescent: quickly fading or disappearing

exude: to ooze or discharge gradually

f.: abbreviation for forma, a secondary taxon rank that denotes a special form of a species or variety (such as a different color or shape)

farinose: covered with flour-like particles

farinaceous: an odd, unpleasant odor similar to that of wheat flour

fascicle: group or bundle

fasciculate: arranged in groups or bundles

fertile surface: layer of tissue on which the spore-bearing structures are produced (= hymenium)

fiber: a hair-like structure

fibril: a short, slender fiber

fibrillose: composed of fibrils

fibrous: tough and stringy

filamentous: composed of filaments

filiform: long and very slender

fissured: split or cracked to forming one or more long, narrow openings

flexuous: bent alternately in opposite directions

floccose, flocculose: with tufts, or small tufts, of loose cottony material

fluted: having longitudinal ridges or ribs

free gills: gills attached only to the cap, not to the stipe

fruitbody: macroscopic sexual reproductive structure produced by certain fungi

fuliginous: smoky or sooty grayish brown, dark and dusky

fulvous: reddish cinnamon color

fungarium: permanent organized repository for preserved fungi

furfuraceous: covered with coarsely granular or bran-like particles; (= scurfy)

furrowed: marked with grooves

fuscous: dark brownish gray to brownish black

gastroid: puffballs and species that look like a bare stomach; they release spores from inside the "stomach"

gelatinous: having a jellylike consistency

genera: plural of genus

genus: formal group of similar, presumably closely related species

glabrous: smooth, naked

gleba: a fleshy, gelatinous, or powdery spore-bearing inner mass

gluten, glutinous: a slimy layer on fungi formed by hyphae that release sticky contents

granulose: covered with small sugar-like particles

gregarious: fruitbodies occurring close together but not fused

hardwood: flowering tree with broad leaves

herbaceous: plants that have little or no woody tissue

herbarium: a permanent organized repository for dried or otherwise preserved plant specimens, often including fungi

hirsute: covered with long, stiff hairs

hispid: covered with stiff hairs or bristles

hoary: having a whitish to grayish sheen, like frost

horny: tough and hard; resembling animal horn

hyaline: transparent and nearly colorless

hygrophanous: color becomes paler (often tan) with loss of water

hymenium, hymenia: the spore-producing, or fertile, tissue of a fruitbody, such as the gills of a mushroom or layer of asci

hypha, hyphae: long, tube-like element(s) that make up the body (mycelium) of a filamentous fungus

hypogeous: fruitbodies that grow below ground; compare epigeous

inamyloid: remaining colorless or yellowish in iodine solutions

incised: as if cut into

inclusions: structures or substances contained within

incrustations: crusts or hard coatings on the surface

incurved: bent or depressed inward

intergrade: two species that may be hybridizing to form fruitbodies with intermediate forms, often observed as intermediate colors

involute: inrolled, especially in reference to the margin of an apothecium or edge of a mushroom cap

KOH: potassium hydroxide, usually used in a 3–5 percent aqueous solution; NaOH (sodium hydroxide) or lye are substitutes

labyrinth: an intricate structure consisting of interconnecting passages; a maze

labyrinthine: relating to or constituting a labyrinth

lacerate, lacerated: cut or torn and sometimes ragged

laciniate: torn or coarsely cut

lacunose: covered with pits or indentations or containing chambers

lignin: a complex organic chemical deposited in the cell walls of many plants, making them rigid and woody

lineate: arranged as or marked with lines or streaks

locule: a cavity, especially one occurring in a stroma within which the hymenium is produced

lubricous: slightly greasy or slippery to the touch

Lugol's reagent: an iodine-containing solution used, both microscopically and macroscopically, to test for color changes; important for Ascomycota

macrofungi: moderately large to large species of fungi

macroscopic, macroscopically: visible with the unaided eye, or at low magnification such as obtained with a 10-power hand lens

marbled: marked with streaks of a different color

margin: the edge of a cap or a rim on a club-shaped stipe

imarginate, non-marginate: with or without a rim on a swollen stipe base

maze: an intricate structure consisting of interconnecting passages; a labyrinth

Melzer's reagent: an iodine-containing solution used, both microscopically and macroscopically, to test for color changes; see amyloid and dextrinoid

micron (μm): microscopic unit of measurement in the metric system equal to 1/1000 mm

microscopic, microscopically: visible only with a microscope

milk: liquid drops that form on gills or a cut on a fungus

mllk mushroom: a fungus in the genus *Lactarius* and one *Suillus* species

mm: unit of measurement in the metric system, equal to 1/10 cm and 1/1000 meter; 1 in. = 25.4 mm

monophyletic: arising from one ancestral group

montane: pertaining to cool, moist mountainous areas below timberline

mottled: marked with spots or smears of one or more different colors

mucilage: a slimy or sticky substance

mucilaginous: slimy or sticky

multicellular: composed of many cells

multichambered: having several to many chambers

multifurcate: branched repeatedly

multiguttulate: having several to many oil drops

multinucleate: having several to many nuclei

mushroom: a loosely defined term for the large, conspicuous fruitbodies of certain fungi, often confined to those with fleshy texture, those with cap and stalk, or those bearing gills

mycelium, mycelia: the body of a filamentous fungus, composed of a network of complexly branched hyphae

mycologist: one who studies fungi

mycology: the study of fungi

mycoparasite: a fungal parasite on another fungus

mycorrhizae: symbiotic associations of fungi and the roots of plants

mycorrhizal: pertaining to a nearly universal mutualistic symbiosis between the hyphae of fungi and the roots of plants

mycosis, mycoses: a fungal infection

mycota: all the species of fungi that inhabit a given area

nodulose: with small bumps, knobby

nonamyloid: remaining colorless or yellowish in Melzer's or Lugol's reagent

non-apiculate: lacking short projections

nonchitinoid: lacking chitin

n.p.: *nomen provisorium*, a Latin term meaning an unpublished or provisional name

obtuse: blunt, not pointed; compare with acute

ocher, ochre: variable reddish, brownish, orangish, yellowish color

ochraceous: having an ochre color

oligotrophic: relatively low in plant nutrients

olivaceous: having an olive-green color

pallid: a very pale shade of any color, almost white

papilla, papillae: a small, nipple-shaped elevation

papillate: having a papilla

parasite: a fungus (or other organism) that obtains its nutrients from another living organism (its host), which suffers a negative impact as a result

parasitized: having one organism living in or on another organism and deriving food from it

partial veil: membrane from cap edge to the stipe

pellicle: A membrane (skin) on the surface of a mushroom cap

perennial: lasting or remaining active for many years

peridial: pertaining to a peridium

peridium: the outermost layer of a fruitbody such as a truffle or puffball

periodole: the tiny "eggs" found in bird's nest fungi that contain and protect the spores

phallic: resembling an erect penis

phylogenetic tree: a branching diagram showing evolutionary relationships

phytopathogenic: having the ability to cause disease in plants

plicate: creased or pleated

polygonal: a two-dimensional shape having many angles

polyhedral: an enclosing three-dimensional shape having only ± flat sides

polymorphic: having several to many shapes

polyphyletic: arising from several ancestral groups

polysaccharide: a complex sugar

pruinose: appearing to be covered with a fine powder

pseudorhiza: a root-like structure joining the fruitbody and mycelium

pubescent: covered with short, soft, fine hairs

pulvinate: cushion-shaped

punctate: covered with tiny scales or spots

pungent: having a strongly sharp taste or smell

pustulate: appearing as though covered with blisters

pustules: pimple-like or blister-like elevations

rays: recurved outer portions of a fruitbody that have split open

recurved: curved backward

reflexed: bent downward and outward

repand: having a margin that is wavy and turned back or elevated

resinous: sticky and soft or hard, like pitch or tar

resupinate: lying flat on the surface

reticulate: having a net-like pattern

reticulate-echinulate: having a net-like pattern and small, pointed spines

reticulum: net-like system of ridges; often in reference to a spore

revivable: able to resume natural shape and functions when placed in moist conditions

rhizome: a horizontal underground stem

rhizomorph: a group of thick, thread-like strands of hyphae growing together as a single organized unit

ribbed: having a pattern of raised bands

riparian: relating to forests or wetlands adjacent to rivers and streams

rugose, rugulose: wrinkled, finely wrinkled

saccate: shaped like a sack

saprotroph: obtains its nutrition by decomposing dead organic material

saprotrophic: pertaining to a saprotroph

scabers: soft, rasp-like scales

scabrous: covered with scales

scale: an erect, flattened, or recurved projection or torn portion of a surface

scalloped: edged with a series of curved projections

scattered (habit): fruitbodies occurring closely enough together to be considered a group, but farther apart than for gregarious

sclerotium, sclerotia: a hard, compact mass of hyphae, and sometimes soil, and from which a fruitbody may arise

scurfy: covered with coarsely granular or bran-like particles (= furfuraceous)

scutellate: shaped like a small shield

section: a subdivision of species within a genus or subgenus with distinguishing features that set them apart from other members of the genus or subgenus

semi-effused: partly adhering to a surface and partly projecting outward

sessile: without any sort of stalk

seta, setae: a usually thick-walled, yellow or brown, bristle or hair-like element

setose: having setae

sigmoid: s-shaped

sinuous: wavy

sinus: a bent surface or curve

solitary (habit): fruitbodies occurring singly without others nearby

sordid: dirty or dingy in appearance

spathulate: shaped like a spoon or spatula, ovate with a narrowed base

species: the lowest category in the classification of organisms (although some species are further divided into subspecies, varieties, or forms)

spermatic odor: the odor of male sperm

spinose, spinulose: covered with spines or small, fine spines

spore: single-, or sometimes multi-celled, sexual or asexual reproductive propagule formed by fungi and plants such as algae, ferns, and mosses

spore print: when a mushroom has been placed spore-bearing surface (often, gills or pores) down on a piece of white paper and left until the spores have dropped onto the paper; used to determine spore color

sporulate: to produce and release spores

spp.: abbreviation for multiple species

squamules: small scales

squamulose: covered with small scales (squamules)

ssp.: abbreviation for subspecies

stature: overall form, such as tall and slender or short and stocky

stellate: star-shaped

sterigma, sterigmata: tiny, slender stalks on which basidiospores are formed

sterile: non-fertile

stipe: the structure that arises from the substrate and supports the head or cap of a fungus, equivalent to stalk of a plant

stipitate: having a stipe (stalk)

striate: having fine parallel or radiating lines or grooves; often in reference to a spore or cap

striations: fine parallel or radiating lines or grooves

stroma, stromata: cushion-like tissue on or in which the spores are produced; fruitbody

stromal: pertaining to a stroma

stromatized: made into a stroma

subacute: somewhat pointed or sharp-edged

subalpine: located on mountain slopes just below the tree line

subgenus: a grouping of mushrooms within a genus, each with a distinguishing feature or feature that sets them apart from other members of the genus; often this feature is the presence or absence of a toxic chemical

subiculum: a more or less dense, felty, or cobwebby mat of hyphae covering the substrate and from which the fruitbodies arise

substrate, substratum, substrata: substance on or within which the vegetative body of the fungus is growing and from which the mushroom arises; usually soil, woody material, or leaf litter, but can include insects or other mushrooms

subterranean: existing underground

subtruncate: appearing somewhat cut off

subturbinate: top-shaped with a somewhat flattened apex

sulcate: having parallel or radial grooves, the depth of which is greater than in striate and less than in plicate

superficial: occurring at or on the surface

symbionts: different organisms that live together

synonym: an alternate, usually less preferred, name for an organism, especially a later or illegitimate name no longer in use (in this book, a synonym is represented by an equal sign, =)

tawny (color): dull yellowish brown, like a lion's coat

teleomorph: the sexual state of a fungus, characterized by the presence of spores formed through meiosis such as ascospores or basidiospores

temperate: relating to or denoting a region or climate characterized by having a mild temperature

terrestrial: growing on the ground

tomentose: covered with long, soft, densely matted hairs; woolly

tomentum: covering of long, soft, matted hairs

troop: a large number of fruitbodies in a dense group

trooping: having large numbers of fruitbodies in a dense group

truncate: ending abruptly, appearing chopped off

tubercle: a small, rounded projection

tuberculate: having tubercles

tuberous: round and swollen

turbinate: top-shaped, like an inverted cone

turgor: rigidity caused by increased absorption of fluid

ubiquitous: widely distributed in nature

umbilicate: having a small central depression, like a navel (an innie)

umbo: a rounded, raised portion in the center of the cap of a mushroom

umbonate: a mushroom with an umbo

undulate, undulating: wavy

undulate-rugose: wavy and wrinkled

unicellular: consisting of a single cell

universal veil: membranous structure enclosing the entire fruitbody when young

unzoned: a mushroom cap lacking alternating circular bands of color

vacuolar: having small cavities

vacuolate: possessing vacuoles

vacuole: a small cavity or space that is often bound by a membrane and filled with fluid

var.: abbreviation for variety

velutinous: covered with short, fine, soft hairs, like velvet

vermiform: worm-shaped

verrucose, verruculose: warty, finely warty

villose: covered with long, thin, straight, non-interwoven hairs

vinaceous: color of red wine spilled on paper

violaceous: having violet tones

viscid: sticky

viscid-gelatinous: sticky and having a jelly-like consistency

zonate/zoned: with alternating circular bands of color

μm (micron): microscopic unit of measurement in the metric system equal to 1/1000 mm

Index of Common Names

Most mushrooms do not have common names, though notable edible and poisonous mushrooms sometimes have numerous common names. And sometimes the same common name can apply to two or more unrelated species or to multiple closely related species.

The following common names are either in widespread use in Cascadia or point to an interesting feature of the mushroom. See MycoMatch for many more common names.

Admirable bolete	*'Boletus'* (*'Aureoboletus'*) *mirabilis*
alder tongues	*Taphrina occidentalis*
amadou	*Fomes fomentarius*
angel wings	*Pleurocybella porrigens*
artist's conk	*Ganoderma applanatum*
barometer earthstar	*Astraeus hygrometricus*
bear's head	*Hericium abietis*
beefsteak false morel	*Gyromitra esculenta*
bird's nests	*Crucibulum crucibuliforme, Cyathus* species, *Nidula* species
bitter boletes	*Caloboletus* species
black chanterelle	*Craterellus calicornucopioides*
blewit	*Lepista nuda*
blue-capped polypore	*Albatrellus flettii*
blue chanterelles	*Polyozellus* species
butter boletes	*Butyriboletus* species
candy cap	*Lactarius rubidus*
cannon fungus	*Sphaerobolus stellatus*
chicken of the woods	*Laetiporus* species

club-footed clitocybe	*Ampulloclitocybe clavipes*
coral mushrooms	*Artomyces, Clavulina, Lentaria, Phaeoclavulina, Ramaria* species
corn smut	*Ustilago maydis*
crested coral	*Clavulina coralloides*
deadly webcap	*Cortinarius rubellus*
death cap	*Amanita phalloides*
deer mushrooms	*Pluteus* species
destroying angel	*Amanita ocreata*
early morel	*Verpa bohemica*
earthballs	*Scleroderma* species
earthstars	*Astraeus, Geastrum* species
elfin saddle	*Helvella vespertina*
ergot	*Claviceps purpurea*
eyelash cups	*Scutellinia* species
fly agaric	*Amanita muscaria*
funeral bells	*Galerina* subgenus *Naucoriopsis*
gemmed amanita	*Amanita gemmata*
golden bolete eater	*Hypomyces chrysospermus, Hypomyces microspermus*
gray-gilled wood lover	*Hypholoma capnoides*
gray morel	*Morchella tomentosa*
gray-veiled amanita	*Amanita porphyria*
green-gill parasol	*Chlorophyllum molybdites*
green-gilled wood lover	*Hypholoma fasciculare*
green stain	*Chlorociboria* species
grisette	*Amanita vaginata*
hawk wing	*Sarcodon imbricatus*
hedgehogs	*Hydnum* species
honey mushroom	*Armillaria* species
Indian paint fungus	*Echinodontium tinctorium*
inky cap	*Coprinopsis atramentaria, C. striata, C. romagnesiana*
king bolete	*Boletus edulis*
kurotake	*Boletopsis grisea*
landscape morel	*Morchella importuna*
late oyster mushroom	*Sarcomyxa serotina*

spring king	*Boletus rex-veris*
strawberries and cream	*Hydnellum peckii*
sweetbread mushroom	*Clitopilus prunulus*
sweet tooth	*Hydnum repandum* group
the prince	*Agaricus augustus*
the walnut	*Gyromitra montana*
tulip cup	*Sarcosphaera coronaria*
tumbling puffball	*Bovista pila*
turkey tail	*Trametes versicolor*
velvet foot	*Flammulina velutipes*
velvet pax	*Tapinella atrotomentosa*
western blond morel	*Morchella tridentina*
western grisette	*Amanita pachycolea*
western half-free morel	*Morchella populiphila*
white chanterelle	*Cantharellus subalbidus*
white matsutake	*Tricholoma murrillianum*
witch's butter	*Tremella mesenterica, mesenterella*
wood ear	*Auricularia americana*
woodland agaricus	*Agaricus abruptibulbus*
woolly chanterelle	*Turbinellus floccosus*
woolly pine spike	*Chroogomphus tomentosus*
yellow chanterelle	*Cantharellus formosus*
yellow morel	*Morchella americana*
yellow-veiled amanita	*Amanita augusta*

Index of Scientific Names

Note: Numbers in bold indicate the page where the main descriptions of the species and genera are found; other references lead to secondary mentions.

Text copyright © 2024 by Michael Beug

Photographs copyright © 2024 by Michael Beug

All rights reserved.

Published in the United States by Ten Speed Press, an imprint of the Crown Publishing Group, a division of Penguin Random House LLC, New York.
TenSpeed.com

Ten Speed Press and the Ten Speed Press colophon are registered trademarks of Penguin Random House LLC.

Originally published in the United States by The FUNGI Press, Batavia, IL, in 2021.

All photographs are by the author with the exception of those listed here:

Harley Barnhart: *Agaricus diminutivus*; *Aleuria aurantia*; *Coltricia cinnamomea*; *Coprinopsis* cf. *romagnesiana*; *Craterellus calicornucopioides*; *Morchella tomentosa*; *Mycena nivicola*; *Pholiota brunnescens*; *Psathyrella epimyces*; *Russula cyanoxantha*; *Trichaptum biforme*

Kit Scates Barnhart: *Agaricus xanthodermus*; *Amanita ocreata*; *Amanita phalloides* (bronze form), *Astraeus pteridis*; *Boletus barrowsii*; *Chlorophyllum molybdites*; *Chroogomphus pseudovinicolor*; *Clavariadelphus caespitosus*; *Clavicorona taxophila*; *Coprinopsis striata*; *Flammulina velutipes*; *Fomitopsis ochracea*; *Gomphidius maculatus*; *Gymnopus aeruginosus*; *Gymnopus fuscopurpureus*; *Humaria hemisphaerica*; *Hydnellum peckii*, bottom image; *Hygrophorus marzuolus*; *Hygrophorus pudorinus*, right image; *Hygrophorus pusillus*; *Hygrophorus sordidus*; *Inocybe geophylla*; *Inocybe napipes*; *Inosperma calamistratum* group; *Irpex lacteus*; *Ischnoderma resinosum*, left image; *Lacrymaria lacrymabunda*; *Lactarius alnicola*; *Lactarius mucidus* var. *mucidus*; *Lactarius resimus*; *Lenzites betulina*, left image; *Leotia lubrica*; *Leptonia fuligineomarginata*; *Leptonia incana*; *Leratiomyces squamosus* var. *squamosus*; *Macrotyphula fistulosa*; *Melanoleuca cognata*; *Multiclavula vernalis*; *Neolentinus ponderosus*; *Phellodon atratus*; *Pholiota carbonaria*; *Pluteus tomentosulus*; *Pycnoporus cinnabarinus*; *Ramaria amyloidea*, left image; *Ramaria formosa*; *Ramaria rubrievanescens*; *Rhodocollybia butyracea*; *Rhodocollybia oregonensis*; *Russula cerolens* left image; *Russula densifolia*, right image; *Russula olivacea*, right image; *Russula vinososordida*; *Russula xerampelina* group, right image; *Sarcodon rimosus*; *Thaxterogaster occidentalis*; *Tricholoma atrofibrillosum*; *Tricholoma ammophilum*; *Xeromphalina brunneola*

Paul Stamets: *Bridgeoporus nobilissimus*, *Psilocybe azurescens*, *Psilocybe allenii*, *Psilocybe baeocystis*

Image on page 24 by fotografiecor; ruler on inside cover by svitlananiko, stock.adobe.com.

Typefaces: Lineto's Bradford, Klim Type Foundry's Pitch Sans, Dinamo's Monument Grotesk

Library of Congress Control Number: 2023033052

Trade Paperback ISBN: 978-19848-6347-8
eBook ISBN: 978-19848-6348-5

Printed in China

Editor: Julie Bennett | Production editor: Ashley Pierce
Designer and Art Director: Annie Marino | Production designers: Mari Gill and Faith Hague
Production manager: Serena Sigona
Copyeditor: Lisa Theobald | Proofreader: Mikayla Butchart
Marketer: Stephanie Davis

10 9 8 7 6 5 4 3 2 1

Second Edition